"This book by Jorge A. Colombo represents a careful update and critical discussion of a theme that underlines the biological conditioning of human physiology and behaviour since early postnatal days. The human gut microbiome constitutes a dynamic, interactive component of human physiology since earlier days of individuals under normal conditions, exacerbated/modified under pathological and inadequate nutritional conditions. It covers fundamental domains of the interactions involving gut microbiome, diet, behaviour, poverty, and child development. The book calls the attention on open issues involving conceptual themes on human neurobiological integration and its impact on crucial developmental and social domains."

Prof. Dr. Andreas Reichenbach, (formerly) Paul Flechsig Institute of
Brain Research, Leipzig University, Germany

"I happily endorse the book by Jorge Colombo who is the most prominent neuroanatomist and behaviorist and writer and painter – the man of Renaissance proportions indeed. I, as well as many people in my field (Neuroglia), are constantly referring to many of his seminal discoveries. I sincerely believe that this new book will be another milestone."

Prof. Dr. Alexei Verkhratsky, professor of Neurophysiology,
The University of Manchester, UK

THE HUMAN BIOME AND HUMAN BEHAVIOUR

The book represents a critical update on interactions between the host and its gut microbiome that conditions the socio-biology of the mind and behaviour.

Evidence has been scientifically approached and reveals that our conscious behaviour involves a complex interplay of multiple non-conscious domains, including complex host-gut microbiome relationships.

The book describes trends and issues on which there is increasing evidence of the impact of host-gut microbiome interactions on behaviour and cultural construction of self-perception. This suggests the need to re-evaluate traditional, basic concepts of human development. Additionally, it calls attention to open issues involving conceptual themes on neurobiological integration and its impact on early developmental and social domains on the typical extended period of human postnatal helplessness during which the basic scaffolding of mental development is completed. It also deals with the impact of poverty and inadequate early feeding habits on individual cognitive development, performance, and social construction. It discusses the need to reformulate views and policies on social marginalisation, child poverty, and malnutrition involving host-gut microbiome imbalances.

The spectrum of possible behaviours in all species and its plasticity depends on an integrated vector of basic components involving the genetic code, social and physical environmental, developmental conditions, the relative condition of dominance or submission in social settings – or prey/predator in the Natural Kingdom – and on its physiological and anatomical construction profiles.

Graduate, postgraduate and teachers interested in areas connected with anthropology, social medicine, early education, and health policymakers will benefit greatly from this book.

Jorge A. Colombo, MD, PhD. is a former Full Professor at the University of South Florida (USA) and Principal Investigator at the National Research Council (CONICET, Argentina). He is also a former Fellow of several international organisations, including NIH, von Humboldt Foundation, DAAD, and British Royal Society.

THE HUMAN BIOME AND HUMAN BEHAVIOUR

A Biopsychological Perspective

Jorge A. Colombo

LONDON AND NEW YORK

Cover image: Getty Images © Dusan Stankovic

First published 2025
by Routledge
4 Park Square, Milton Park, Abingdon, Oxon OX14 4RN

and by Routledge
605 Third Avenue, New York, NY 10158

Routledge is an imprint of the Taylor & Francis Group, an informa business

© 2025 Jorge A. Colombo

British Library Cataloguing in Publication Data
A catalogue record for this book is available from the British Library

Library of Congress Cataloging-in-Publication Data
A catalog record has been requested for this book

ISBN: 9781032698397 (hbk)
ISBN: 9781032678511 (pbk)
ISBN: 9781032698380 (ebk)

DOI: 10.4324/9781032698380

Typeset in Galliard
by Taylor & Francis Books

With love, to my wife, Beatriz, to the memory of my parents and my brother, Emilio, to my children and grandchildren

To the memory of my friends and mentors in the field of neurosciences, Prof. Dr. Enrique T. Segura (Argentina), Charles Sawyer (USA), and Karl Zilles (Germany)

CONTENTS

FIGURES

FOREWORD

As stated by Kandel et al. (2013), *all behaviour is an expression of neural activity.* Beyond its integrative concept – on physiological domains, *the final common pathway* – the above statement implies a universe of biological and behavioural interactive components that modulate or condition such neural activity, in essence, the socio-biology of the mind. In this regard, from an evolutive point of view, it seems appropriate to include here a new view on the possible neuronal origin – a crucial event in neurobehavioural evolution –, according to Najle et al. (2023),

> The assembly of the neuronal and other major cell type programs occurred early in animal evolution... Our findings indicate that key neuronal developmental and effector gene modules evolved before the advent of cnidarian/bilaterian neurons in the context of paracrine cell signaling.

The aforementioned diverse components and conditioners of manifest behaviour in the individual may or may not reach conscious levels. However, these have been scientifically analysed and revealed that our conscious behaviour is – to a significant extent – the result of a complex interaction of multiple non-conscious domains. This book aims to focus on one such domain, reflected in the following paragraphs.

> Animals and human beings have evolved in intimate and constant association with a complex microflora and microfauna. Under natural conditions, the development and functions of their tissues are influenced by countless microorganisms which are always present in the digestive and respiratory tracts, and probably also in other organs. It is to be expected, therefore, that anatomical structures and physiological needs have been determined in part by the microbiota which prevailed during evolutionary development, and that many manifestations of the body at any given time are influenced by the microbiota now present. It can be taken for granted, in other words, that the microbiota is part of the environment to which animals and man have had to become adapted, and on which they have come to depend.
>
> *(Dubos et al. 1965)*

> Bacteria came before us and most likely will remain for some time after our extinction. Meantime, they thrive on us and send us continuous signals with which the immune system keeps them at bay. Man, the most intelligent creature on Earth, usually disregard bacteria. But they often show their enormous potency and kill the host as a reminder of how frail our life is on Earth, where man is only a transient guest and not the master.
>
> *(Bocci 1992)*

> [As Lewis Thomas (1974:142) commented when considering self and symbiosis: "This is, when you think about it, really amazing. The whole dear notion of one's own Self—marvelous, old free-willed, free-enterprising, autonomous, independent, isolated island of a Self—is a myth."]
>
> *(Gilbert et al. 2012)*

Beyond reports as those stated above and of Savage (1977), new developments on the complexity of organisms and bacterial interactions suggest the need to re-evaluate traditional, basic concepts in the field of human biological construction and self-perception. That is, concepts involved with constructing an integrated vision of the physiology of complex organisms, with particular reference to humans. An updated map of the human gut microbiome was provided by Leviatan et al. (2022).

*

   At the public level, human self-perception is usually biased by a concept of species dominance that tend to underplay or disregard our shared construction with the ecosystem and our symbiotic nature with comparatively simpler microorganisms that predated by aeons the emergence of complex organisms, and more specifically of *Homo* species.

   Hence, this book does not intend to provide a complete review of the field of host-microbiome interactions but rather describe trends and issues on which there is increasing evidence regarding several aspects of its behavioural consequences and the cultural construction of our self-perception. It intends to focus on open issues involving conceptual themes of human neurobiological integration and its impact on crucial developmental and social domains. The account provided here stresses such conceptual need and projects a more realistic, holistic vision of our species' biological nature as an emerging and interactive event within the biological world. This last view impacts the consequences of collective actions of our species in the world habitat and the need to reformulate views and policies regarding the ecosystem in which all species, including humans, thrive.

   A holistic consideration of the evolutive process in the Natural Kingdom cannot avoid integrating the microbial species of early appearance on evolutive domains, nor the interactions among them and with the later evolving macrospecies. Considering that the Natural Kingdom is populated by the fittest and most adaptable organisms to environmental demands, convivial and sharing survival strategies are constantly subjected to *trial and error*. Under this concept, ultimate dominance does not necessarily equate to circumstantial, physical predominance but to species survival and its emerging behavioural profiles. Hence, the commensalisms and competence across animal and plant species represent an everlasting, interactive play of trials and errors where survival is the only valid unit of measure.

How does the emergence of the human species fit in this integrated, dynamic context? In the Natural Kingdom, the probability of survival and the opportunity to develop an optimal performance of an individual will depend on their innate and acquired abilities to ensure survival of their genetic composition. Among the latter are those that depend on the quality and duration of their social support and upbringing, a characteristic of the species adapted to the ecological niche.

After millennia, the "Darwinian jungle" changed actors, but survival conditions persist in its essence. Given the fragility and plasticity of the child's brain and the helplessness of the infant, the risks associated with their rearing remain in force related either to conditions of poverty or subjected to human cultural variables that act as conditioners of its mental development and freedom of thought. In it, their upbringing is a modulated behavioural characteristic by cultural dynamics. This, as well as parenting, is significantly affected by strategies and decisions related to public policies, often resulting in marginalisation, exposure to unsanitary conditions, lack of biomedical containment and education, and food deficiency that would compromise the emergence of a healthy interactive host-gut microbiome interaction. These conditions enclose the risk of an immoral counter-evolutionary condition to the extent that it violates one of the basic premises that allowed the survival of our species: an upbringing according to the requirements posed by the full development of a demanding neonate unable to achieve self-sufficiency for a long postnatal period. More yet, considering that during the first years of life, the brain consumes more than 50% of the whole body's energy and about 80% during the first months of extrauterine life (see later in the text). The issue of feeding, providing nutritional quality, and the healthy development of a gut microbiome are closely linked to the optimal survival and growth of the human neonate.

The natural rearing period depends on vulnerability and postnatal self-sufficiency, the duration and conditions of which are highly variable in the evolutionary tree. These survival conditioners are coupled with nutritional factors that assure a healthy gut microbiome development and its interaction with the host. It may be worth insisting that, compared to other species of the Natural Kingdom, humans have a very extended period of postnatal helplessness during which basic scaffolding is completed of their brain and mental development. Therefore, the necessary rearing period-imposed conditions are particularly demanding for the emergence of *Homo sapiens*. Based on paleoanthropological data, even in related species of the genus *Homo* or of the other primates, the period of immaturity was shorter, just as the skull laxity was more limited in postnatal brain growth. The result was an earlier childhood maturation and a smaller brain volume, therefore not equivalent to the *sapiens* child's. What conditioning of community behaviours at the dawn of *sapiens* would have resulted from such containment demands in a world with other insecurities? If these demands were not effectively satisfied, the survival of *Homo sapiens* would have collapsed before attaining its total capacity. Perhaps we should extract social ethics teachings from this.

Territoriality and prevalence are universal domains in survival behaviours within the Natural Kingdom. Whether solitary or gregarious social habits, animals and plants show territorial behaviours towards conspecifics and prey or reject those competing with feeding or reproductive resources. *Homo sapiens* carry such a backpack, and its culturally transformed – or hidden – expression takes place as dominant behaviour and hierarchical social constructs, either spontaneously, under social dynamic circumstances, or transformed into virtual (cultural, ideological) domains. Its bonded relationship should be added with the continuously evolving sophisticated material culture, which interactively evolves our collective mind and virtual constructions. *Homo sapiens*' evolution is wrapped around the construction of instruments of progressive complexity and power, developed into a cultural–material technology that resets the relationships among individuals and the environment.

Through time, based on social repression or "socialisation", cultural strata of variable "thickness" have been constructed on top of drives implicit in our animal condition. Nevertheless, it failed in their deactivation or suppression and only succeeded in reformulating or transitorily repressing them, as the history of human civilisation demonstrates deviant behaviours expressed at the individual and collective levels. Interaction with the physical and cultural environments continues to model our ethnic variations. However, our primary organisation is bound to ancestral demands that imprinted basic drives: childcare, territorialism, reproduction, survival, secure feeding sources, dominance, and accumulative behavioural experience. Their expression, affected by changed environmental conditions – physical and sociocultural – poses the probability of continuous friction between the neurobiological and cultural *tectonic plates*. Indeed, the plasticity of our brain and mind construction (depending on cultural issues and contexts) provides for adaptative responses. However, these have not cancelled the framework of primary drives and biological construction imprinted in the heart of our animal inheritance but instead affect the probability or sociocultural profile expression. Bio-social interaction continues to model social, behavioural trends, or social phenotypes on top of the primary, deeply entrenched survival and prevalent drives conformed according to the basic structure of the ancestral animal nature. This has conditioned the disparate social, cultural, and cognitive conditions among nations, ethnicities, and individuals that have contributed to building our modern world's fractured composition.

Sociocultural differences between global communities and, within them, among their constituents result from a different history and dynamics of genomic-environmental interactive human constructions and justify limiting the current concept of globalisation to limited socio-economic domains. However, on the backstage of human behaviours lies a primal condition related to developing a healthy host-gut microbiome interaction. That is, as the interaction of two different organismic domains, the host and a complex set of microorganisms.

In this regard, which and how much of our current behaviours – individually and as a global community – are driven by ancestral, inherited traits imprinted in our animal condition and by the components of our host-gut microbiome interactions? The cultural domain implied in this question pertains to our identity and pertinence to bio-social constructions and ecological interaction.

*

The human sense of pride and dominant macrospecies has set up a risky, if not dreadful, set of events that affects its own present and future living conditions as a global community. It also has a widespread impact on ecological and survival conditions for all living species. In an unstable equilibrium among living creatures, prey and predator behaviours built during aeons a renewable series of actions and species in an interactive system with ecological variables. In essence, the multiple trials of emerging interactive species through aeons of terrestrial and marine life have not vanished one key behavioural component of survival – dominance (Colombo 2022).

Occasionally, geological and environmental events placed organisms at a critical level of subsistence derived from many species' exhaustive physiological adaptability ranges. However, no species has the means to alter the biological equilibrium within the living creatures of the Natural Kingdom and the ecosystem to the extent that humans do. Once humans took control of the natural range as property, they disturbed the biological and climate equilibrium, triggering species devastation and climate excesses. No other macrospecies could challenge its power and stop its actions besides silent – microbial – partners sharing the destructive path that *Homo sapiens* took. This placed our culturally and technologically developing civilisation in a dominant position that impacted the construction of our relationship with the ecosystem and its living organisms. In turn, it generated a sort of *megalomaniac species* that displaced our biological reality and the web of mutual dependencies.

> It is too easy to argue that since we are the only hominin species left on the planet we must be unique and special in some respect. This proposition does not tell us what were the paths that our ancestors took to become so distinctive and to what extent we share partially, or entirely, this supposed uniqueness with our present or past relatives.
>
> *(d'Errico and Stringer 2011)*

Besides affecting the species' equilibrium and creating climate imbalances among living creatures, *Homo sapiens* reigned with no opposition among macrospecies. However, this description results in an incomplete biological and ecological equilibrium picture. Original, ancient, ubiquitous, simple microorganisms that sparked the long-life chain of animal and plant evolution (Archaea, Bacteria, and Eukaryotes), with aeons of adaptive interactions and survival, did not abandon their *front-line* position in biological evolution regardless of successively changing dominant macrospecies.

> Although invisible to the naked eye, prokaryotes are an essential component of the earth's biota. They catalyse unique and indispensable transformations in the biogeochemical cycles of the biosphere, produce important components of the earth's atmosphere, and represent a large portion of life's genetic diversity.
>
> *(Whitman et al. 1998)*

Recent data suggest that the human body is not so exclusively human after all. Specifically, humans share their bodies with approximately 10 trillion microorganisms, collectively known as the microbiome. Chief among these microbes are

> bacteria, and there is a growing consensus that they are critical to virtually all
> facets of normative functioning.
>
> *(Smith and Wissel 2019)*

As stated by Stilling et al. (2014), the tight association of the human body with trillions of colonising microbes is the result of a long evolutionary history. This book will consider this theme and project onto human socio-behavioural developmental domains. McFall-Ngaia et al. (2013) called attention to current-day relationships of protists with bacteria, from predation to obligate and beneficial symbiosis that were probably already operating when animal species first appeared. These ancient eukaryote–bacterial interactions provide insights into metazoan evolution, from the origins of complex multicellularity to the drivers of morphological complexity.

*

> …a journalist approached the renowned psychologist Donald Hebb and asked for
> his opinion on which factor contributed more to the development of personality,
> nature or nurture. Hebb responded that to pose this question was akin to asking
> what contributed more to the area of a rectangle, the length or the width.
>
> *(Meaney 2001)*

This book has a dual objective. While it does not intend to provide an exhaustive review of the field of bacteriology related to human health domains, it circles the evolutive concept and involved variables on the construction of human behavioural phenotypic profiles and their interactive consequences. It involves integrating the microbiome concept with neuro-behavioural domains within current habits. Additionally, this analysis will include human actions that raise further concern about the impact of poverty and inadequate early feeding habits on individual cognitive development, performance, and social construction. It compounds the host-gut microbiome interactions and individual outcomes.

Maintaining a beneficial microbiota requires a homeostatic equilibrium within microbial communities and between the microorganisms and the intestinal interface of the host (Sommer et al. 2017). It should be considered that changes in the composition of the gut microbiota (dysbiosis) may be associated with several clinical conditions, including obesity and metabolic diseases, autoimmune diseases and allergies (Goulet 2015). However, clinical disease developments here would only be considered if they provide an additional perspective on human behavioural performance. Initially, special consideration will be given to primal living forms since they constitute an evolutive link and share the physiology of current complex organisms, as mentioned in the quote by Smith and Wissel (2019).

Public educational policies aimed at promoting/enforcing solidarity, knowledge and the practice of imagination and divergent thinking or conceptual expansion could only attain their objectives if they are integrated with equity in social construction. Hence, social marginalisation, child poverty and malnutrition involving gut microbiome imbalance are their significant opponents. Furthermore, they are at the base of unethical, counter-evolutive, regressive engines (Colombo 2010, 2013, 2015).

In essence, the primaeval human host-gut microbiome – an "ancient marriage" – is indissoluble and resistant to any divorce attempt. Hence, humankind's worthwhile survival – if it includes creativity and solidarity domains – requires an ever-improving happy marriage between host and gut microbiome demands, a condition fully unmet among vast populations.

Microbial colonisation of mammals is an evolution-driven process that modulates host physiology... Our results suggest that during evolution, the colonisation of gut microbiota has become integrated into the programming of brain development, affecting motor control and anxiety-like behavior.

(Heijtz et al. 2011)

References

Bocci, Velio. "The Neglected Organ: Bacterial Flora has a Crucial Immunostimulatory Role." *Perspectives in Biology and Medicine*, vol. 35, no. 2, 1992, pp. 251–260, doi:10.1353/pbm.1992.0004.

Colombo, Jorge A. *Somos La Especie Equivocada?: Una Mirada Evolutiva Sobre Las Responsabilidades De Nuestra Especie*. Buenos Aires: Eudeba, 2010.

Colombo, Jorge A. *Bajo Libertad Condicionada ¿Hacia La Conquista De Grados De Libertad?* Buenos Aires: Imago Mundi, 2013.

Colombo, Jorge A. *Los Homo Sabios*. Buenos Aires: Buenos Aires Books, 2015.

Colombo, Jorge A. *Dominance Behavior: An Evolutive and Comparative Perspective*. Cham: Springer International Publishing, 2022.

Crane, Leah. "Asteroid Samples Contain a Building Block of RNA." *New Scientist*, vol. 257, no. 3432, 2023, p. 19, doi:10.1016/s0262-4079(23)00560-2.

d'Errico, Francesco, and Chris B. Stringer. "Evolution, Revolution or Saltation Scenario for the Emergence of Modern Cultures?" *Philosophical Transactions of the Royal Society B: Biological Sciences*, vol. 366, no. 1567, 2011, pp. 1060–1069, doi:10.1098/rstb.2010.0340.

Dubos, René, *et al.* "Indigenous, Normal, and Autochthonous Flora of the Gastrointestinal Tract." *The Journal of Experimental Medicine*, vol. 122, no. 1, 1965, pp. 67–76, doi:10.1084/jem.122.1.67.

Gilbert, Scott F., *et al.* "A Symbiotic View of Life: We Have Never Been Individuals." *The Quarterly Review of Biology*, vol. 87, no. 4, pp. 325–341, doi:10.1086/668166.

Goulet, Olivier. "Potential Role of the Intestinal Microbiota in Programming Health and Disease: Figure 1." *Nutrition Reviews*, vol. 73, no. suppl. 1, 2015, pp. 32–40. doi:10.1093/nutrit/nuv039.

Heijtz, R. D., *et al.* "Normal Gut Microbiota Modulates Brain Development and Behavior." *Proceedings of the National Academy of Sciences*, vol. 108, no. 7, 2011, pp. 3047–3052, doi:10.1073/pnas.1010529108.

Kandel, Eric R. *Principles of Neural Science*. 5th ed., New York; Toronto: McGraw-Hill Medical, 2013.

Kobayashi, Kensei, *et al.* "Formation of Amino Acids and Carboxylic Acids in Weakly Reducing Planetary Atmospheres by Solar Energetic Particles from the Young Sun." *Life*, vol. 13, no. 5, 2023, p. 1103, www.mdpi.com/2075-1729/13/5/1103, doi:10.3390/life13051103.

Leviatan, Sigal, *et al.* "An Expanded Reference Map of the Human Gut Microbiome Reveals Hundreds of Previously Unknown Species." *Nature Communications*, vol. 13, no. 1, 2022, doi:10.1038/s41467-022-31502-1.

McFall-Ngai, Margaret, *et al.* "Animals in a Bacterial World, a New Imperative for the Life Sciences." *Proceedings of the National Academy of Sciences*, vol. 110, no. 9, 2013, pp. 3229–3236, doi:10.1073/pnas.1218525110.

Meaney, Michael J. "Nature, Nurture, and the Disunity of Knowledge." *Annals of the New York Academy of Sciences*, vol. 935, no. 1, 2001, pp. 50–61, doi:10.1111/j.1749-6632.2001.tb03470.x.

Najle, Sebastián R., *et al.* "Stepwise Emergence of the Neuronal Gene Expression Program in Early Animal Evolution." *Cell*, vol 186, no. 21, 2023, pp. 4676–4693, doi:10.1016/j.cell.2023.08.027.

Savage, D. C. "Microbial Ecology of the Gastrointestinal Tract." *Annual Review of Microbiology*, vol. 31, 1977, pp. 107–133, www.ncbi.nlm.nih.gov/pubmed/334036, doi:10.1146/annurev.mi.31.100177.000543.

Smith, Leigh K., and Emily F. Wissel. "Microbes and the Mind: How Bacteria Shape Affect, Neurological Processes, Cognition, Social Relationships, Development, and Pathology." *Perspectives on Psychological Science*, vol. 14, no. 3, 2019, pp. 397–418, doi:10.1177/1745691618809379.

Sommer, Felix, *et al.* "The Resilience of the Intestinal Microbiota Influences Health and Disease." *Nature Reviews Microbiology*, vol. 15, no. 10, 2017, pp. 630–638, doi:10.1038/nrmicro.2017.58.

Stilling, Roman M., *et al.* "Friends with Social Benefits: Host-Microbe Interactions as a Driver of Brain Evolution and Development?" *Frontiers in Cellular and Infection Microbiology*, vol. 4, no. 147, 2014, doi:10.3389/fcimb.2014.00147.

Whitman, W. B., *et al.* "Prokaryotes: The Unseen Majority." *Proceedings of the National Academy of Sciences*, vol. 95, no. 12, 1998, pp. 6578–6583, www.cen.ulaval.ca/merge/pdf/Whitman1998.pdf, doi:10.1073/pnas.95.12.6578.

ACKNOWLEDGEMENT

To Flavia C. Abdullah for language and format editing.

PRELIMINARY QUOTES ON THE HYPOTHESIS REGARDING THE PRIMAEVAL ORIGIN OF BIOORGANIC MOLECULES, A BASIS FOR PREBIOTIC EVOLUTION

Samples from the asteroid *Ryugu* contain uracil, one of the four building blocks of RNA, as well as niacin and other compounds that are important for living organisms. This lends credence to the idea that the ingredients for life were brought to Earth by space rocks. Japan's *Hayabusa 2* spacecraft returned 5.4 grams of asteroid dust from *Ryugu* at the end of 2020. Now an analysis of samples has revealed uracil and complex organic molecules.

(Crane 2023)

Since the energy flux of space weather, which generated frequent *solar energy particles* (SEP) from the young Sun in the first 600 million years after the birth of the solar system, was expected to be much greater than that of galactic cosmic rays, we conclude that SEP-driven energetic protons are the most promising energy sources for the prebiotic production of bioorganic compounds in the atmosphere of the Early Earth.

(Kobayashi et al. 2023) (Text in italics inserted by the author JAC)

1

EVOLUTIVE CONSIDERATIONS

The person who has undoubtedly contributed the most to our knowledge about the role of symbiotic bacteria in development is Margaret McFall-Ngai (see McFall-Ngai 2002; McFall-Ngai, Henderson and Ruby 2005). Her work has proved crucial to the adoption of an ecological perspective in developmental biology, that is, a perspective that acknowledges the importance of environmental factors on development (Gilbert 2001, 2002, 2005; Gilbert and Epel 2009).

(Pradeu 2011)

The overall pattern from studies of the core is that we share a functional core microbiome (the collection of genes represented in the microbiota), but not a core microbiota.

(Lozupone et al. 2012)

The analysis of the complex, slow process of the evolution of the human species belongs to one of the most extraordinary scientific achievements. However, it is very daunting that nature slides past most public opinion because of culturally alleged magic or concealed forces that would be beyond human comprehension. This divergence provided grounds for stories and representations adapted to cultural and environmental domains and traditional profiles that moulded human beliefs and social drives.

Let us consider that, most notably, life evolution is a complex, conditional process that evolved during aeons under diverse environmental demands, involving a constant dynamic interaction. During such an interactive process, an unknown number of species failed to survive, and newcomers populated vacant niches, triggering new evolutionary chains in the Natural Kingdom.

*

DOI: 10.4324/9781032698380-1

The spectrum of possible behaviours in all species and their plasticity depends on an integrated vector of essential components. These are defined by the genetic code, social and physical environment, developmental conditions, the relative condition of dominance or submission in social settings, or prey/predator in the Natural Kingdom, and on its physiological and anatomical construction profiles. Much later, *Homo sapiens* incorporated new variables to this complex equation based on cultural and technological opportunities to expand cognitive potential and mental awareness, innovate social construction, increase species dominance, and expand environmental abuse.

However, this view requires the inclusion of an additional dimension related to ancient, primal, individual, or group commensalism and sharing with a universe of prokaryotes integrated into an individual's animal and plant physiologies. These considerations are usually a hidden component in our assessment of our complex human construction and will be the primary thrust of the present book, with its social dominance consequences.

The measure of a species' dominance success could be analysed using several vectors. On natural grounds, these include its ability or capacity to survive or outlive other species on a limited period if we look at the survival of other *Homo* species, its capacity to exert dominance within the Natural Kingdom, and invade natural domains based on its creations, many of which could survive their biological creators. Not all these survival vectors have identical life spans. While humans may excel in the latter two domains, the biologically simplest species that sparked complex forms of life since the dawn of Earth has shown as a biological domain their ubiquitous and adaptive resources that out-measure any other living species.

*

Aeons past, early Earth conditions were practically devoid of atmospheric oxygen. When our planet formed about 4.5 billion years ago, it was a sterile ball of flaming rock, slammed by meteorites and carpeted with erupting volcanoes. Within a billion years, microorganisms inhabited it (possible sources mentioned before). *Simple* forms of life – whether generated on our planet or because of "pollution" from outer space – would have colonised Earth probably more than 3.0 billion years ago, able to cope with harsh environmental conditions. Only anaerobic organisms could develop under those initial conditions, implying a hard limitation for growing organisms of greater biological complexity. Thus, the biological domain until then was populated by extremophile and anaerobic organisms until dominant physical conditions of early planetary conditions eased up and became amenable to initial complex forms of life.

This process involved microorganisms, cyanobacteria capable of producing photosynthesis, which progressively transformed an oxygen-devoid atmosphere into an oxygenated one. These established the basic conditions for developing species requiring oxygen consumption. This includes extinct – former dominant species – and species we share with the world today following a series of mass extinctions. The evolution of simple forms showed that while archaea and bacteria were devoid of nuclei, eukaryotes included all past and present complex life forms with cell nuclei enclosing their DNA.

As mentioned by Dance (2021),

Archaea are more than just oddball lifeforms that thrive in unusual places — they turn out to be quite widespread. Moreover, they might hold the key to understanding how complex life evolved on Earth. Many scientists suspect that an ancient archaeon gave rise to the group of organisms known as eukaryotes, which include amoebae, mushrooms, plants and people – although it's also possible that both eukaryotes and archaea arose from some more distant common ancestor.

*

Davidov and Jurkevitch (2009) posed that since prokaryotes cannot perform phagocytosis, how the endosymbiont entered its host is an enigma. The authors considered that accumulating data suggests that the eukaryotic cell originated from a merger of two prokaryotes, an archaeal host and a bacterial endosymbiont, and suggest that a predatory or parasitic interaction between prokaryotes would provide a reasonable explanation.

Douglas (2014) provided a succinct evolutive perspective on eukaryotes,

> Various resident microorganisms provide a metabolic capability absent from the host, resulting in increased ecological amplitude and often evolutionary diversification of the host. Some microorganisms confer primary metabolic pathways, such as photosynthesis and cellulose degradation, and others expand the repertoire of secondary metabolism, including the synthesis of toxins that confer protection against natural enemies.

All extant eukaryotes have mitochondria or are descendants of mitochondrial ancestors, raising the possibility that the acquisition of mitochondria defined the evolutionary origin of eukaryotes, as discussed by Embley and Martin (2006),

> Mitochondria in previously unknown biochemical manifestations appear to be universal among eukaryotes, modifying views about the nature of the earliest eukaryotic cells and testifying to the importance of endosymbiosis in eukaryotic evolution.

Iyer et al. (2004) analysed the issue of genes found only in animals and bacteria. They considered that it is consistent with a late gene transfer from a bacterial source into eukaryotes after separating the animal and fungal lineages, an early process, as also stated by Hoffmeister and Martin (2003). This supports the concept that animals have carried resident microorganisms since at least the emergence of sponges, as quoted by Blaser and Falkow (2009). As stated by Embley and Martin (2006), the persistence of mitochondrially derived organelles in all eukaryotes and plastids in some lineages provides phylogeny-independent evidence for the occurrence of those symbiotic events. Additionally, the fact that all extant eukaryotes have mitochondria or are descendants of mitochondrial ancestors raises the possibility that this cellular organelle defined the evolutionary origin of eukaryotes (Embley and Martin 2006).

As suggested by Zilber-Rosenberg and Rosenberg (2008), eukaryotes probably arose from prokaryotes and remained associated since then as a *holobiont (hologenome)*, considered to be an animal or plant with all its associated microorganisms as a unit of selection in evolution. Furthermore, as mentioned by these authors, prokaryotes would have remained in a close relationship with eukaryotes since then (cf., Hickman 2005), whether on the surfaces of animals and plants or able to grow inside animal or plant cells, as endosymbionts. According to the far-reaching conceptual synthesis by Gilbert et al. (2012),

> Only with the emergence of ecology in the second half of the 19th century did organic systems...complement the individual-based conceptions of the life sciences.... They have not only revealed a microbial world of much deeper diversity than previously imagined, but also a world of complex and intermingled relationships – not only among microbes, but also between microscopic and macroscopic life (Gordon 2012). These discoveries have profoundly challenged the generally accepted view of individuals.

According to Gilbert et al. (2012), the diversity of symbionts is functional in completing metabolic pathways and serving other physiological functions. Symbionts constitute a second mode of genetic inheritance, providing selectable genetic variation for natural selection. Since their origins, eukaryotes and their microbial symbionts have ranged from obligate intracellular to extracellular microbes that forge mutualistic, commensal, and parasitic interactions (Stilling et al. 2014). Lombardo (2007) has referred to this "access to mutualistic endosymbiotic microbes" as a driving force in the evolution of animal sociality. Thus, Gilbert et al. (2012) further supported Zilber-Rosenberg and Rosenberg's (2008) concept that animals are not individuals by the traditional anatomical, physiological, immunological, genetic or developmental accounts. Instead, developmental symbiosis generates *holobionts*, i.e., organisms composed of numerous genetic lineages, the interactions of which are crucial for their development and maintenance, subject to environmental cues. A concept that, according to Roughgarden et al. (2017), consists of a host and diverse microbial symbionts and functions as distinct biological entities anatomically, metabolically, immunologically, and developmentally. As mentioned in Münger et al. (2018) – quoting Lloyd (2017) – , because the microbial component in the holobiont can vary faster than that of the host, microbial diversity may enhance the holobiont's adaptation when selection occurs under fluctuating environmental conditions. Carrier and Reitzel (2017) further considered that studies of the host-associated microbial repertoire are useful in highlighting the importance of microbes in host biology and ecology. This concept would aid in understanding how it may have driven or directed the evolutionary trajectory of the holobiont.

In this regard, Bordenstein and Theis (2015) commented that animals and plants should no longer be viewed as autonomous entities but rather as "holobionts" composed of the host plus all its symbiotic microbes. As stated by these authors, the hologenome concept is a holistic view of genetics in which animals and plants are polygenomic entities; thus, variation in the hologenome can lead to phenotypes upon which natural selection or genetic drift can operate. These authors provided a list of holobionts principles and their hologenomes as units of biological organisation. Among them, it seems proper to underline that complex multicellular eukaryotes are not and have never been autonomous organisms. Instead, they are biological units organised from numerous microbial symbionts and their genomes, and holobionts and their hologenomes do not change the rules of evolutionary biology. Although these concepts redefine what constitutes an individual animal or plant, they do not represent a fundamental rewriting of Darwin's and Wallace's theory of evolutionary biology. Bordenstein and Theis (2015) finally stated that,

> From a specific standpoint, the holobiont and hologenome concepts redefine that which constitutes an individual animal or plant by asserting that hosts and their symbiotic microbes are complex units of biological organisation upon which ecology and evolution can act. From a general standpoint, the concepts assert that macrobe (*macrobiotic*) evolution has been driven by both population and community genetics and that symbiotic microbes and nuclear genes hold equivalent significance in the origin of new holobiont species.
>
> *(Word in italics added by JAC)*

And as McFall-Ngaia et al. (2013) further stated,

> Woese and George Fox opened a new research frontier by producing sequence-based measures of phylogenic relationships, revealing the deep evolutionary history shared by all living organisms. This game-changing advance catalysed a rapid development and application of molecular sequencing technologies, which allowed biologists for the first time to recognise the true diversity, ubiquity, and functional capacity of microorganisms. This recognition, in turn, has led to a new understanding of the biology of plants and animals, one that reflects strong interdependencies that exist between these complex multicellular organisms and their associated microbes.

Theis et al. (2016) asserted that the identity of the term holobiont is based not only on recognising hosts and their obligate symbionts but also on emphasising the diversity and complexity of facultative symbionts and their dynamic associations within a host. According to these authors, microbial genomes can be stable or labile components of the hologenome and can be vertically or horizontally transmitted, and the traits that they encode are context-dependent. These authors draw a conceptual parallelism between holobionts and hologenomes with genomes and chromosomes in that they all reflect different levels of biological information. The terms holobiont and hologenome would imply structural definitions, though their utility remains subject to debate. Furthermore, hologenomic variation may arise by mutation and recombination in the host and microbiome and by acquiring new microbial strains from the environment, changing microbial abundance, and horizontal gene transfer among microbes. The hologenome is an entity that embraces the coevolutionary processes inherent in much of macroscopic biology (Theis et al. 2016). In further discussing the propriety of the term hologenome, these authors proposed the term "phylosymbiosis" to describe the concordance between a host phylogeny (evolutionary relationships) and microbial community dendrogram (ecological relationships) based on the degree of shared taxonomy and abundance of members of the community. These authors considered the hologenome concept a comprehensive and relevant eco-evolutionary framework for which critical questions remain. Yet, to Prescott et al. (2016),

… life can be viewed through the lens of the holobiont, that is, the multicellular eukaryote and the inseparable colonies of persistent symbionts which together form a critically important unit of anatomy, physiology, immunology, growth, and evolution. At the very least, the host plus its microbiome (microbiota and their collective genomes) can be regarded as an ecological community.

*

As mentioned by Walter and Ley (2011), the three-way interactions between human genetics, diet, and microbiota fundamentally shaped modern populations and continue to affect health globally. Foster and Bell (2012) considered that the typical result of adaptation to other microbial species would be competitive rather than cooperative phenotypes. They stressed that studying these interactions between bacterial strains will benefit from a paradigm based on competitive, not cooperative, evolution. On the contrary, Rakoff-Nahoum et al. (2016) considered that understanding whether microbial communities are formally cooperative is central to predicting their evolutionary and ecological stability, involving microbial dynamics that affect the balance of cooperation and competition within and between species. The authors claimed to have found evidence of strong eco-evolutionary interactions within the microbiota that are likely to be central to the functioning of these complex communities. Oliveira et al. (2014) stressed that in these microbial communities, ecological interactions are usually common and strong due to high cell density and that microbes possess many phenotypes that influence the reproduction and survival of surrounding cells and function to support cells of the same genotype. Schluter et al. (2015) considered adhesion a crucial process in microbial competition and forming surface-attached microbial communities. According to these authors, a genotype can dominate a community simply by being more adhesive than its competitors. Zhao et al. (2019) considered the case of intra-individual *Bacteroides fragilis* populations containing substantial *de novo* nucleotide and mobile elements diversity, which would preserve years of within-person history. Additionally, results would demonstrate that *B. fragilis* adapts within individual microbiomes, pointing to factors that promote long-term gut colonisation.

Doolittle and Booth (2017) considered that emphasis should be placed on the biochemical functions of the microbiota and contributions to host biology, which are more conserved than the holobiont's microbial community. These authors propose casting metabolic and developmental interaction patterns as units of selection rather than the taxa responsible for them. Huitzil et al. (2018) supported the basic concept that bacteria have strongly influenced multicellular organisms' evolution and biological functions. Based on complex interdependencies between microbes and multicellular organisms, their conceptual construction as a holobiont, i.e., their function as a biological unit, provides a conceptual synthesis. However, these authors concede that whether holobiont is or is not an evolutionary unit is still a matter of debate. They further add that,

Understanding why microbial diversity is necessary for the evolution and adaptation of the host, and why disease arises when such diversity is lost, is a fundamental question with still no definitive answer.

The taxonomic composition of the microbiota varies among individuals and displays a range of sometimes redundant functions that modify the physicochemical environment of the host and may alter selection pressures, according to Suzuki and Ley (2020). Thus, host-associated microbiotas would have affected the ecology and evolution of various organisms, including humans, with the microbiota potentially affecting host evolution by modifying the adaptive landscape. This would have altered selection pressures on host phenotypes without affecting coevolution.

Moran and Sloan (2015) raised some preconditions to the functional acceptance of the concept of the hologenome as the primary unit of selection based on the still unknown mechanisms to limit the tendency of microorganisms to evolve selfish traits at the expense of fitness of the holobiont. More recently, on evolutionary domains, Moran et al. (2019) posed that,

> ... a central distinction is that gut microorganisms may matter only because hosts have evolved dependence on their presence (evolutionary "addiction") (Douglas 2010; Moran 2002); ... these would confer fitness benefits though they don't extend the original ecological range of the host. Alternatively, gut microorganisms may confer new capabilities entirely lacking in the ancestral host, and thus expand their host's ecological range and evolutionary success.

Thus, formal conceptual definitions of the *holobiont* and *holobiome* may remain debatable. However, their implicit functional interaction provides a framework to analyse further host-microbiome interactions and their impact on our understanding and perception of animal species' biological identity and evolution, including humans.

*

Our species would not have existed if a series of complex physical and chemical circumstances in primaeval Earth conditions had not occurred since "extremophile" microorganisms – capable of proliferating in the most extreme physical or chemical conditions – and cyanobacteria would not have colonised the planet in early, remote, times (the Archean and Proterozoic aeons, of the Precambrian period over 3,000 million years ago). Thus, the Precambrian period established the initial biological conditions for future developments. This early dominance by simple microorganisms provided the basic condition for a subsequent chain of plant and animal species changing dominances within the Natural Kingdom. As stated by Stanley (1973),

> According to modern ecological theory, high diversity at any trophic level of a community is possible only under the influence of cropping. Until herbivores evolved, single-celled algae of the Precambrian were resource-limited, and a small number of species saturated aquatic environments. In the near absence of vacant niches, life diversified slowly. Because the changes required to produce the first algae-eating heterotrophs were therefore delayed, the entire system was self-limiting. When the "heterotroph barrier" was finally crossed in the late Precambrian, herbivorous and carnivorous protists arose almost simultaneously, for no major biological differences separate the two groups.

The explosive radiation of life in the late Precambrian was produced by a kind of self-propagating mutual feedback system of diversification between trophic levels, which was initiated by the advent of heterotrophy. The nearly simultaneous origin and explosive radiation of heterotrophic protists, metaphytes, and metazoans were inevitable. In contrast, the self-limiting, saturated autotrophic system of the earlier Precambrian had strongly inhibited diversification, to delay crossing of the "heterotroph barrier" for hundreds of millions of years.

(Author comment. Concept of autotrophy: species that synthesise their own food through the process of photosynthesis)

Over aeons, severe weather events, astronomical impacts, extinctions of life forms, competition for habitat, and geophysical events extinguished original, complex forms of life and created opportunities to emerge new species in different ecological niches. While some still survive, others seemed destined to endure but were cancelled by unexpected events or self-limited survival adaptability. Today, our human species – exerting its biological and ecological dominance – is an agent of the extinction of species and native ecosystems in what has been marked as the Anthropocene epoch.

Nevertheless, ancient, primal, abundant, microscopic commensals and cohabitants of animal inner (gut microbiome) and outer environments have been a meaningful component of species evolution, decay, and behavioural construction. These were associated with animal evolution in all its forms, sticking to life forms and contributing to their eventual survival success or failure. As noted in Münger et al. (2018), different and characteristic gut microbiome communities may have coevolved with carnivores, herbivores and omnivorous species. Even more, disruptions of our microbial communities have been implicated in various diseases, including mental health. Additionally, the microbiome affects the environment, too, since imbalances in microbial communities can lead to zones of low oxygen in our oceans – fish die-offs – while agricultural depletion of bacteria can lead to barren soils.

*

The role of dominance in human behaviour and social construction was earlier analysed (Colombo 2022) as related to within-species conflicts, macrobiotic species' conviviality with other species, and the biodiversity within a dynamic ecosystem. The basic concept to be considered here is that Natural History evolved as a multifactorial, multi-dependent interaction among macro and micro species. In an interactive equilibrium, mutual dependence condition involves the chain of species and the containing ecosystem in an unstable equilibrium. Significant perturbation at any of these levels will severely affect such equilibrium and degradation of living species. In this domain, human actions have placed climate conditions at risk, other species' survival and quality of life involving their survival instinct and survival capacity.

Biological dominance in Natural History could be approached in terms of predominating species during their actual lifetime – on metaphorical terms, *those that may win the battle but lose the war* – or on species that endure longer, even beyond their natural predators. After all, survival and adaptation are basic vectors of biological life. The instability of the implied equilibrium and range of conditions for species adaptations constitute a major event that our era has been labelled Anthropocene since humans have set new environmental menaces to living conditions. While some losses could be restored with the corresponding changes in how humans interact with the ecosystem, a significant loss of species variety appears to be definitive. The former requires a major conceptual change regarding human presence, actions, and means of living that affect current dominant corporative and political profiles and goals.

*

Human existence entered the Natural Kingdom following a series of aleatoric geological and environmental events and a by-product of animal species flourishment and decay in an interactive prey-predator and survival context. Under human dominance, terrestrial dominance ceased to be within local contexts to become global and a factor of widespread ecological impact with unforeseeable consequences. However, underneath and imbricated with animal and plant species, bacteria and other unicellular entities were the grandparents of all and will probably be the bereaved of all species.

As a species becomes more specialised, it also loses degrees of freedom, thus becoming more fragile. On collective grounds, this equation involves a third component for humans, expressed as an uneven cumulative cognition coupled with creativity and technological developments. This has consequences on global living conditions, conforming to a fractured reality of human civilisation.

Additionally, humanity has become a means for species mass extinction and severe climate changes. While the human population already exceed 8 billion people, the rest of living beings and climate conditions that conform to the ecosystem are gradually changing – degrading – under human actions, promoting an accelerated, progressive disappearance of species. In this domain, humans have already destroyed two-thirds of the world's tropical forests and half of the coral reefs, and millions of species are at risk of being forgotten, as many others have disappeared before the Anthropocene. This era represents a geological and cultural concept proposed to characterise the impact of human action on the environment. Within this domain falls a needed answer to the following question: which and how much of our current drives – individually and as a global community – are driven by ancestral, inherited dominant traits imprinted in our animal condition?

Regarding the previous question, it falls within the universal domain of species plasticity. As species gain specificity, they become less adaptable to environmental variable demands. While the order of bacterial and similar simple organisms cope with it by quickly adjusting their chemical and reproductive domains, complex, multisystemic organisms either fall to their limited adaptive plasticity or develop non-natural strategies – as technology – to cope with them, as is the case of some human populations; reaching strategies or solutions that are not of universal access and applicability in due form and time.

Thus, the overall ecologic domain points towards two possible developing dominant biological universes. The primal, ancient microbial population has been tested under innumerable environmentally changing conditions since life emerged on Earth, and humans are aiming at technologically solving its mis-adaptations and changing the ecological system. With one mandatory imposition, i.e., whatever we do and wherever humans go, we are built to carry and share with our bacterial partners and keep them in a stable equilibrium with our biological system. Otherwise, our regulatory physiological system enters into disarray, with a foreseeable and unwanted spectrum of possible outcomes.

Perhaps we should be conscious that *Homo sapiens* represents *another brick in the wall* of Natural History. Humans would appear to have not yet assumed the implications of such conceptually appalling history projected into our species' construction; it defines the limits and conditions of our freedom to change as a collective society.

References

Blaser, Martin J., and Stanley Falkow. "What are the Consequences of the Disappearing Human Microbiota?" *Nature Reviews Microbiology*, vol. 7, no. 12, 2009, pp. 887–894, doi:10.1038/nrmicro2245.

Bordenstein, Seth R., and Kevin R. Theis. "Host Biology in Light of the Microbiome: Ten Principles of Holobionts and Hologenomes." *PLoS Biology*, vol. 13, no. 8, 2015, doi:10.1371/journal.pbio.1002226.

Carrier, Tyler J., and Adam M. Reitzel. "The Hologenome Across Environments and the Implications of a Host-associated Microbial Repertoire." *Frontiers in Microbiology*, vol. 8, 2017, doi:10.3389/fmicb.2017.00802.

Colombo, Jorge A. *Dominance Behavior: An Evolutive and Comparative Perspective*. Cham: Springer International Publishing, 2022.

Dance, Amber. "The Mysterious Microbes That Gave Rise to Complex Life." *Nature*, vol. 593, no. 7859, 2021, pp. 328–330, www.nature.com/articles/d41586-021-01316-0, doi:10.1038/d41586-021-01316-0.

Davidov, Yaacov, and Edouard Jurkevitch. "Predation between Prokaryotes and the Origin of Eukaryotes." *BioEssays*, vol. 31, no. 7, 2009, pp. 748–757, doi:10.1002/bies.200900018.

Doolittle, W. Ford, and Austin Booth. "It's the Song, Not the Singer: An Exploration of Holobiosis and Evolutionary Theory." *Biology & Philosophy*, vol. 32, no. 1, 2016, pp. 5–24, doi:10.1007/s10539-016-9542-2.

Douglas, Angela E. "Symbiosis as a General Principle in Eukaryotic Evolution." *Cold Spring Harbor Perspectives in Biology*, vol. 6, no. 2, 2014, pp. a016113–a016113, cshperspectives.cshlp.org/content/6/2/a016113.full, doi:10.1101/cshperspect.a016113.

Embley, T. Martin, and William Martin. "Eukaryotic Evolution, Changes and Challenges." *Nature*, vol. 440, no. 7084, 2006, pp. 623–630, www.nature.com/articles/nature04546, doi:10.1038/nature04546.

Foster, Kevin R., and Thomas Bell. "Competition, Not Cooperation, Dominates Interactions among Culturable Microbial Species." *Current Biology*, vol. 22, no. 19, 2012, pp. 1845–1850, doi:10.1016/j.cub.2012.08.005.

Gilbert, Scott F., et al. "A Symbiotic View of Life: We Have Never Been Individuals." *The Quarterly Review of Biology*, vol. 87, no. 4, 2012, pp. 325–341, doi:10.1086/668166.

Hoffmeister, Meike, and William Martin. "Interspecific Evolution: Microbial Symbiosis, Endosymbiosis and Gene Transfer." *Environmental Microbiology*, vol. 5, no. 8, 2003, pp. 641–649, doi:10.1046/j.1462-2920.2003.00454.x.

Huitzil, Saúl, et al. "Modeling the Role of the Microbiome in Evolution." *Frontiers in Physiology*, vol. 9, 2018, doi:10.3389/fphys.2018.01836.

Iyer, Lakshminarayan M., et al. "Evolution of Cell–Cell Signaling in Animals: Did Late Horizontal Gene Transfer from Bacteria Have a Role?" *Trends in Genetics*, vol. 20, no. 7, 2004, pp. 292–299, doi:10.1016/j.tig.2004.05.007.

Lombardo, Michael P. "Access to Mutualistic Endosymbiotic Microbes: An Underappreciated Benefit of Group Living." *Behavioral Ecology and Sociobiology*, vol. 62, no. 4, 2007, pp. 479–497, doi:10.1007/s00265-007-0428-9.

Lozupone, Catherine A., *et al.* "Diversity, Stability and Resilience of the Human Gut Microbiota." *Nature*, vol. 489, no. 7415, 2012, pp. 220–230, www.ncbi.nlm.nih.gov/pmc/articles/PMC3577372/, doi:10.1038/nature11550.

McFall-Ngai, Margaret, *et al.* "Animals in a Bacterial World, a New Imperative for the Life Sciences." *Proceedings of the National Academy of Sciences*, vol. 110, no. 9, 2013, pp. 3229–3236, doi:10.1073/pnas.1218525110.

Moran, Nancy A., and Daniel B. Sloan. "The Hologenome Concept: Helpful or Hollow?" *PLOS Biology*, vol. 13, no. 12, 2015, p. e1002311, doi:10.1371/journal.pbio.1002311.

Moran, Nancy A., *et al.* "Evolutionary and Ecological Consequences of Gut Microbial Communities." *Annual Review of Ecology, Evolution, and Systematics*, vol. 50, no. 1, 2019, pp. 451–475, doi:10.1146/annurev-ecolsys-110617-062453.

Münger, Emmanuelle, *et al.* "Reciprocal Interactions between Gut Microbiota and Host Social Behavior." *Frontiers in Integrative Neuroscience*, vol. 12, 2018, doi:10.3389/fnint.2018.00021.

Oliveira, Nuno M., *et al.* "Evolutionary Limits to Cooperation in Microbial Communities." *Proceedings of the National Academy of Sciences of the United States of America*, vol. 111, no. 50, 2014, pp. 17941–17946, doi:10.1073/pnas.1412673111.

Pradeu, Thomas. "A Mixed Self: The Role of Symbiosis in Development." *Biological Theory*, vol. 6, no. 1, 2011, pp. 80–88, doi:10.1007/s13752-011-0011-5.

Prescott, Susan L., *et al.* "Biodiversity, the Human Microbiome and Mental Health: Moving toward a New Clinical Ecology for the 21st Century?" *International Journal of Biodiversity*, vol. 2016, 2016, pp. 1–18, doi:10.1155/2016/2718275.

Rakoff-Nahoum, Seth, *et al.* "The Evolution of Cooperation within the Gut Microbiota." *Nature*, vol. 533, no. 7602, 2016, pp. 255–259, doi:10.1038/nature17626.

Roughgarden, Joan, *et al.* "Holobionts as Units of Selection and a Model of Their Population Dynamics and Evolution." *Biological Theory*, vol. 13, no. 1, 2017, pp. 44–65, doi:10.1007/s13752-017-0287-1.

Schluter, Jonas, *et al.* "Adhesion as a Weapon in Microbial Competition." *The ISME Journal*, vol. 9, no. 1, 2015, pp. 139–149, www.nature.com/articles/ismej2014174, doi:10.1038/ismej.2014.174.

Stanley, Steven M. "An Ecological Theory for the Sudden Origin of Multicellular Life in the Late Precambrian." *Proceedings of the National Academy of Sciences*, vol. 70, no. 5, 1973, pp. 1486–1489, doi:10.1073/pnas.70.5.1486.

Stilling, Roman M., *et al.* "Friends with Social Benefits: Host-Microbe Interactions as a Driver of Brain Evolution and Development?" *Frontiers in Cellular and Infection Microbiology*, vol. 4, no. 147, 2014, doi:10.3389/fcimb.2014.00147.

Suzuki, Taichi A., and Ruth E. Ley. "The Role of the Microbiota in Human Genetic Adaptation." *Science*, vol. 370, no. 6521, 2020, p. eaaz6827, doi:10.1126/science.aaz6827.

Theis, Kevin R., *et al.* "Getting the Hologenome Concept Right: An Eco-Evolutionary Framework for Hosts and Their Microbiomes." *MSystems*, vol. 1, no. 2, 2016, doi:10.1128/msystems.00028-16.

Walter, Jens, and Ruth Ley. "The Human Gut Microbiome: Ecology and Recent Evolutionary Changes." *Annual Review of Microbiology*, vol. 65, no. 1, 2011, pp. 411–429, doi:10.1146/annurev-micro-090110-102830.

Zhao, Shijie, *et al.* "Adaptive Evolution within Gut Microbiomes of Healthy People." *Cell Host & Microbe*, vol. 25, no. 5, 2019, pp. 656–667, doi:10.1016/j.chom.2019.03.007.

Zilber-Rosenberg, Ilana, and Eugene Rosenberg. "Role of Microorganisms in the Evolution of Animals and Plants: The Hologenome Theory of Evolution." *FEMS Microbiology Reviews*, vol. 32, no. 5, 2008, pp. 723–735, doi:10.1111/j.1574-6976.2008.00123.x.

2

POSSIBLE LIFE ORIGIN

*

One of the main advances in biology in the 20th century was to highlight the unity of the living world: all currently living beings are made up of the same macromolecules, use the same protein synthesis machinery and the same genetic code, suggesting that they all descend from a common line of ancestors. This does not imply, however, that life appeared only once on our planet: indeed, the question of the single or multiple origin of life, as well as that of the mechanisms of its appearance, are still much debated. However, if other forms of life appeared on Earth, they left no descendants, and were all eliminated by the one we know.

(Forterre et al. 2005)

Nitrogen is a critical ingredient of complex biological molecules. Here we show that nitrogen fixation in the early terrestrial atmosphere can be explained by frequent and powerful coronal mass ejection events from the young Sun—so-called superflares.

(Airapetian et al. 2016)

As stated in previous paragraphs, the origin of "life" represented by biological entities able to adapt, survive and reproduce on Earth generated diverse – perhaps confluent – hypotheses.

Earth underwent a series of geological periods labelled as shown in Figure 2.1.[1,2] According to the *Encyclopaedia Britannica* and the *National Institutes of Health*, bacteria fossils emerged during the Palaeozoic Era, in the Devonian period (see Graph in Figure 2.1). However, there is evidence that their presence would be as remote as about 3.5 billion years during the Precambrian time, including the Archean aeon. Due to the atmospheric condition of planet Earth at that time, they would have been anaerobic. Modern, autotrophic bacteria – organisms that produce complex organic compounds using carbon from simple substances such as carbon dioxide, generally

DOI: 10.4324/9781032698380-2

using energy from light or inorganic chemical reactions – would have appeared approximately 3.8 billion years ago, as Mojzsis et al. (1996) reported. The first organic molecules would have been probably active during the primitive stages of Earth's evolution, combining electric/solar charges in mixtures of Co2, N2 H2, H2S, and CO. (*This comment does not erase other factors as molecular/microbial seeding from outer space, as quoted at the beginning of this book.*)

As summarized in Woese and Fox (1977),

> A phylogenetic analysis based upon ribosomal RNA sequence characterization reveals that living systems represent one of three aboriginal lines of descent: (i) the eubacteria, comprising all typical bacteria; (ii) the archaebacteria, containing methanogenic bacteria; and (iii) the urkaryotes, now represented in the cytoplasmic component of eukaryotic cells.
>
> *(Note. According to this author, in one sense the major ancestors of eukaryotic cells might appropriately be called urkaryotes)*

*

The Primal, Highly Adaptive, Opportunistic Commensal Living Partners

The three domain life forms (see Figure 2.2) – archaea, eukaryotes, and bacteria – would have developed from an original *common or universal ancestor*. As posed by Woese (1998), whatever it was, the ribosomal RNA cryptic entity had spawned three remarkably different primary groupings of organisms (domains), and these necessarily reflected the ancestor's nature,

> The (*common*) ancestor cannot have been a particular organism, a single organismal lineage. It was communal, a loosely knit, diverse conglomeration of

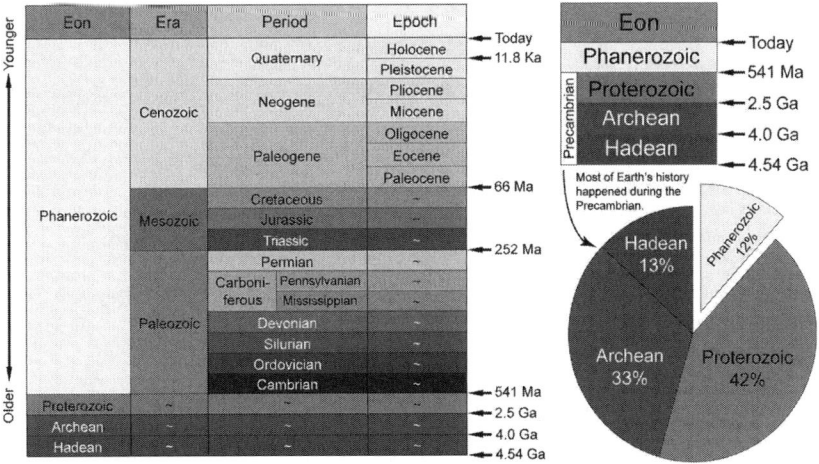

FIGURE 2.1 Geologic Time Scale according to J.R Hendricks for the Earth@Home project.[3]

primitive cells that evolved as a unit, and it eventually developed to a stage where it broke into several distinct communities, which in their turn become the three primary lines of descent... The universal ancestor is not an entity, not a thing. It is a process characteristic of a particular evolutionary stage... The universal ancestor does have an evolutionary history, but that history is physical, not genealogical.

(Woese (1998) (Word in italics added by the author)

This common ancestor was labelled as ***progenote*** by Woese (1987), who described it as,

... a highly unique entity, unlike any life found today. The progenote is a theoretical construct, an entity that, by definition, has a rudimentary, imprecise linkage between its genotype and phenotype.

As mentioned by Woese (1987), this would be the stage of "nucleic acid life" in which translation as we know it has not yet evolved and nucleic acids have both genetic and enzymatic functions.

*

As stated, bacteria arose about 3.8 billion years ago, according to the isotopic results of Mojzsis et al. (1996), and the eukaryotic lineage, which includes humans, probably arose after the oxygenation of Earth's atmosphere 2.2–2.4 billion years ago as based on Kappler et al. (2005). With archaea, protists, and fungi, bacteria remained free-living single cells, although some would become host-associated (Dominguez-Bello et al. 2019). The microbiome has shaped phenotypes in our ancestral lineages by coevolving with the host.

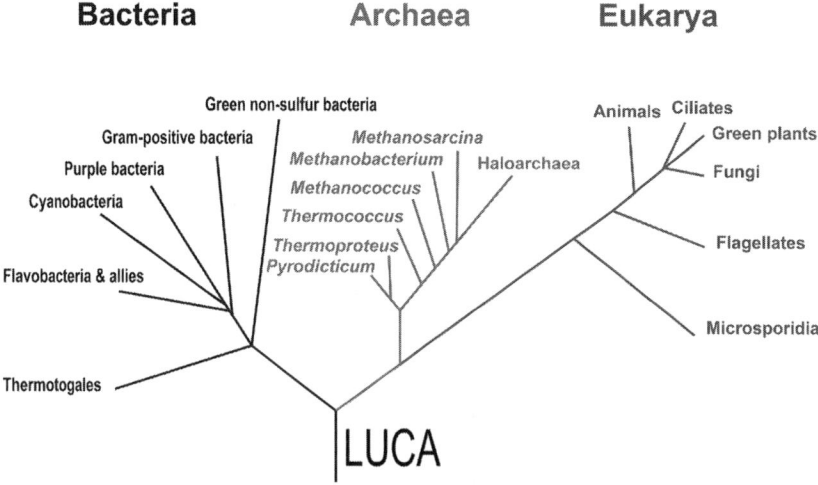

FIGURE 2.2 A phylogenetic tree based on rRNA data, emphasising the separation of bacteria, archaea, and eukarya as proposed by Carl Woese et al. in 1990, with the hypothetical last universal common ancestor (LUCA).[4]

According to Edgell and Doolittle (1997), the most profound phylogenetic division in the universal tree is separating bacteria from the clade comprising archaea and eukaryotes.

Figure 2.2 from *Encyclopaedia Britannica* shows the three basic original life forms represented by archaea, bacteria, and eukaryotes. The following comment (Dance 2021) perhaps characterises the actual situation regarding the origin of primitive cells,

> Archaea are more than just oddball lifeforms that thrive in unusual places – they turn out to be quite widespread. Moreover, they might hold the key to understanding how complex life evolved on Earth. Many scientists suspect that an ancient archaeon gave rise to the group of organisms known as eukaryotes… – although it's also possible that both eukaryotes and archaea arose from some more distant common ancestor.[5]

Additionally, Snel et al. (2001) estimated that the ancestor of the Proteobacteria contained around 2500 genes, and the ancestor of the archaea around 2050 genes. Although it is necessary to invoke horizontal gene transfer to explain the content of present-day genomes, gene loss, gene genesis, and simple vertical inheritance are quantitatively the most dominant processes in shaping the genome. Archeon *Prometheoarchaeum syntrophicum*, considered the ancestor of complex cells, was found at the bottom of the ocean (see Figure 2.3).

Butterfield (2014) offered an analysis of the early evolution of eukaryotes, though considering that there is little consensus as to when or over what timescale it occurred (see also Porter 2020). As the author stated,

> …the assembly of a modern-style eukaryotic cell appears to have taken a billion, maybe a billion and a half years. Last Eukaryotic Common Ancestor clearly opened up new evolutionary opportunities, but diversification and ecological penetration remained decidedly modest through the Mesoproterozoic and early Neoproterozoic – a further billion or so years.
> … the late Palaeoproterozic, and stem-group eukaryotes extend back to the early Archaean. Despite their relatively early establishment, crown-eukaryotes appear not to have become ecologically significant until the middle Neoproterozoic (*). I argue that this billion-year delay was due to the singular, contingent evolution of crown-group animals and their unique capacity to drive co-evolutionary change. Crown-group eukaryotes (Last Eukaryotic Common Ancestor plus all of its descendants) are united by a unique combination of genetic and cytological machinery, yielding a grade of organization fundamentally more complex than found in any prokaryotes (bacteria plus non-eukaryotic archaea). The most conspicuous feature of the eukaryote is their intracellular compartmentalization, supported by an elaborate signaling/ trafficking system and dynamic cytoskeleton.
> *(* Author's note, see Figure 2.1)*

Brocks et al. (2017) considered that the emergence of algae (probably 659–645 million years ago) created more efficient nutrient and energy-transfer food sources that tended to replace cyanobacteria.

The transition from dominant bacterial to eukaryotic marine primary productivity was one of the most profound ecological revolutions in the Earth's history, reorganizing the distribution of carbon and nutrients in the water column.

(Brocks et al. 2017)

Later, Brocks et al. (2023) reported the discovery of abundant proto-steroids in sedimentary rocks of mid-Proteozoic age (see Figure 2.1). According to the authors, these primordial compounds of early protosterol[6]-producing bacteria and stem-group eukaryotes were primarily present in aquatic environments from at least 1,640 to around 800 million years ago, probably involved with the emergence of ancient protosterol-producing bacteria and deep-branching stem-group eukaryotes. Sterols are known to regulate dynamics to maintain membranes in a microfluid state where they can convey critical biological processes (Dufourc 2008).

According to Brocks et al. (2023),

Modern eukaryotes started to appear in the Tonian (stretch) period (1,000 to 720 million years ago), fuelled by the proliferation of red algae (rhodophytes) by around 800 million years ago. This "Tonian transformation" emerges as one of the most profound ecological turning points in the Earth's history.

The Neoproterozoic era – 541–1000 million years ago – transitioned from a primarily bacterial world to the emergence of multicellular grazers, suspension feeders, and predators (Brocks et al. 2017).

As mentioned in Thompson et al. (2017), Earth microbial community composition has been shown to change across gradients of environment, geographic distance, salinity, temperature, oxygen, nutrients, pH, day length, and biotic factors. As posed by Brunet and King (2017), over 600 million years ago, animals would have evolved from a unicellular or colonial organism whose cell(s) captured bacteria with a collar complex, a flagellum surrounded by a microvillar collar. In most animals, multicellularity is established in each generation through serial divisions of a single founding cell, the zygote. It involves a process of cell differentiation under fine spatiotemporal control that delineates the division of labour between the final cell types. According to these authors, the discontinuous phylogenetic distribution of multicellularity and differences in cellular mechanisms argue that multicellularity would have evolved independently in at least 16 different eukaryotic lineages, including animals, plants, and fungi.

In early times, gene transfer (see Figure 2.4) took place as vertical (transmission from parent to offspring) and horizontal or lateral transfer (movement of genetic material between unicellular and/or multicellular organisms not by vertical DNA transmission), considered one of the leading evolutionary forces at work in prokaryotic evolution. This would have allowed the acquisition of functions even from distantly related organisms to adapt to changing environments or colonize new ones.

It is now clear that lateral transfer is far more widespread than had previously been appreciated, and that episodes of rapid evolution (high evolutionary temperature) have been common throughout evolution.[4]

(Woese 1998)

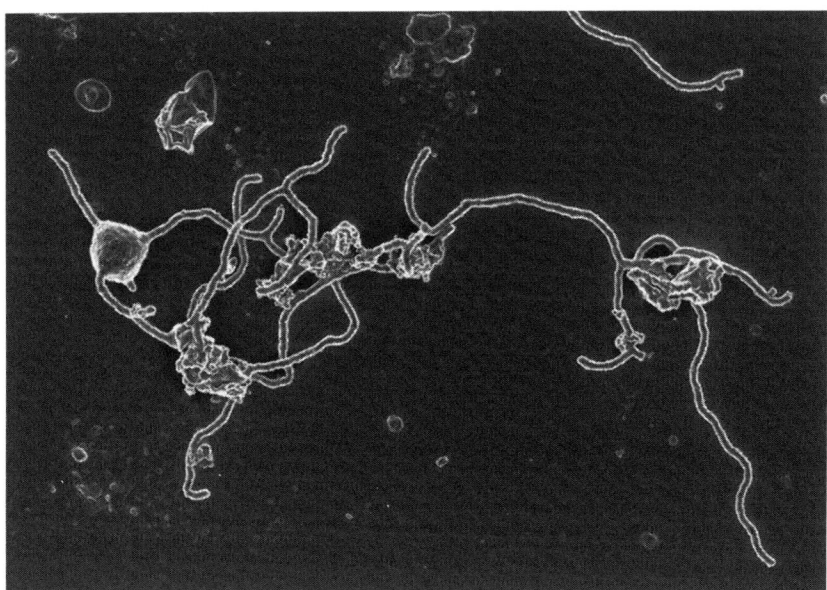

FIGURE 2.3 Scanning electron microscopy image of the Archaeon *Prometheoarchaeum syntrophicum*, found at the bottom of the ocean. It is closely linked to more complex life and is thought to be like the ancestor of complex cells.
Source: Hiroyuki Imachi, Masaru K. Nobu and JAMSTEC.

However, the critical characteristic of the emergence of life would have required self-replication, which would have happened under the catalytic capacity of RNA with self-replicating RNA molecules. Additionally, the physical properties of early cell membranes most probably differed from the contemporary phospholipid-based cell membranes. According to Mansy et al. (2008), the permeability properties of pre-biotically plausible membranes suggest that primitive protocells could have acquired complex nutrients from their environment in the absence of any macromolecular transport machinery; that is, they could have been obligate *heterotrophs* (an organism that cannot manufacture its food by *carbon fixation* and therefore derives its intake of nutrition from other sources of organic carbon, mainly plant or animal matter). According to Koonin (2009), the precellular stage of biological evolution unravelled within networks of inorganic compartments that harboured a diverse mix of virus-like genetic elements. Under the viral model of precellular evolution, the critical components of cells originated as components of virus-like entities. The two surviving types of cellular life forms, archaea and bacteria, might have emerged from the Last Universal Common Ancestor (LUCA), with numerous forms now extinct.

According to Biello (2012), in an article from *Scientific American*, the first cells – or at least the ones that left descendants still extant – would have started in geothermal pools, like those seen at Yellowstone National Park and other geologic hot spots. According to the author, the argument rests on one indisputable observation – enzymes common to all archaea and bacteria are built from potassium, phosphorus, or zinc, not sodium. According to Biello (2012), modern archaea and bacteria possess

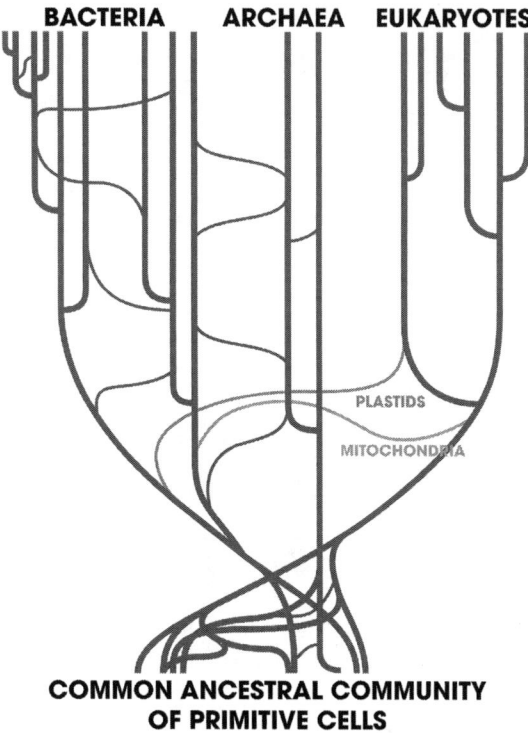

FIGURE 2.4 Horizontal genetic transfer among Bacteria and Archaea, vertical transfer among Eukariotes. By Andrew Z. Colvin - Barth F. Smets, Tamar Barkay (September 2005).[7]

internal fluid low in sodium, and enzymes built from other elements hint that they arose in an environment rich in such elements and relatively sodium-free. Within this debatable issue, this hypothesis is confronted by authors such as Martin and Russell (2003), who propose that life evolved in structured iron monosulphide precipitates in a seepage site hydrothermal mound at a redox, pH and temperature gradient between sulphide-rich hydrothermal fluid and iron-containing waters of the *Hadean Ocean*[1] floor.

The next stage in cell evolution was building up conditions for metabolic energy essentially based on glycolysis, implying the anaerobic breakdown of glucose to lactic acid, providing energy gain, and photosynthesis, allowing the use of solar energy (photosynthetic bacteria). This process released oxygen, which drastically modified in due time the oxidative metabolism, thus setting up the basis for the evolution of new life forms. Many developmental requirements for multicellular organisation, including cell adhesion, cell-cell communication and coordination, and programmed cell death, likely existed in ancestral unicellular organisms (Grosberg and Strathmann 2007).

Multicellular organisms evolved departing from unicellular eukaryotes some 2.0 billion years ago. This schematic, overlooking the early stages of the origin of natural life, provides data on the ancient emergence of prokaryotic unicellular organisms during comparatively Earth's early times, probably more than 3.0 billion years ago.

Regarding biological dominance, if the biological chain is maintained, and if by dominant we define that organism that constitutes the relative mega-predator within the limits of a given ecosystem, then all organisms – whether macro- or microscopic – could claim to be so. In truth, if we were to define it by its permanence, voracity, and adaptability, the kingdom of bacteria would be the major and the most ancestral one since, over time, they would also have superseded mega macropredators.

According to Ley et al. (2006),

> Microbial diversity on our planet is vast: although 55 divisions (deep evolutionary lineages) of Bacteria and 13 divisions of Archaea have been described to date (DeLong and Pace, 2001; Rappe and Giovannoni, 2003; Rondon et al., 1999), much diversity remains unexplored.

Whitman et al. (1998) estimated the number and total carbon of prokaryotes on Earth and concluded that most of them are in seawater, soil, and soil subsurface, and their total carbon would represent about 60–100% of the total carbon found in plants. They further state that previous estimates of prokaryotes representing about one-half of the "living protoplasm" on Earth are microbial may underestimate the protoplasmic biomass of prokaryotes.

*

In constructing progressively complex cells, the respiratory chain in microspecies was provided by the symbiotic inclusion of bacteria. This became the mitochondrial intracellular organelle that generate most of the chemical energy needed to power the cell's biochemical reactions.

> The discovery of mitochondrial and plastid genomes revived the 19th-century view of plastids as domesticated bacterial endosymbionts (Mereschowsky 1910). However, it was not until the more recent developments in molecular biology that the theory really gained ground (Margulis 1970, 1981). The similarities between bacterial respiration and mitochondrial functions as well as the similarities between cyanobacterial photosynthesis and chloroplast functions in plants taken together with molecular phylogenies have provided convincing evidence of the bacterial origins of these modern cellular organelles (Grey 1992) …However, the outstanding unanswered question is the identity of the cell that served as the host for the endosymbiont that eventually evolved into the mitochondrion.
>
> *(Andersson et al. 2003)*

According to Cavalier-Smith (2006), mitochondria evolved from endosymbiotic (i.e., internalisation of a prokaryote by an ancestral eukaryotic cell) purple non-Sulphur bacteria, though the process would still be debatable. Thus, bacteria and chloroplasts became intracellular symbionts, cellular organelles. The serial endosymbiosis theory proposes that a primitive eukaryote successively took up bacteria and blue-green algae to yield mitochondria and chloroplasts, respectively. As mentioned by Keeling and Palmer (2008), the origin of the cytoskeleton and endocytosis enabled eukaryotes to readily engulf and feed on other cells, which are occasionally retained as endosymbionts. Extant eukaryotes arose by horizontal (or lateral) gene transfer. It refers to the movement of genetic information across normal mating barriers between related organisms and thus stands in distinction to vertical transmission of genes from parent to offspring. As reported by Douglas (2014), the evolutionary history of the eukaryotic acquisition of oxygenic photosynthesis by symbiosis with the cyanobacterial ancestor of chloroplasts, which, unlike mitochondria, evolved just once, chloroplasts would have evolved from two different cyanobacterial groups (Keeling 2013).

Needed energy is stored in adenosine triphosphate (ATP) molecules. On evolutionary grounds, Embley and Martin (2006) proposed that all extant eukaryotes have mitochondria or are descended from mitochondrial ancestors, raising the possibility that the acquisition of mitochondria defined the evolutionary origin of eukaryotes. As later stated by Douglas (2014), eukaryotes' evolutionary and ecological success is founded on acquiring aerobic respiration and photosynthesis by symbiosis, with the transition of bacterial symbionts to organelles. This implies that this propensity to form persistent associations has ancient evolutionary roots. Furthermore, according to Douglas (2014), there is increasing evidence of health benefits from the symbiosis between host and microbiome. This would be related to microbial modulation of the signalling networks that coordinate eukaryote hosts' growth and physiological functions.

Without attempting to force parallel, simple comparisons with complex behaviours, some bacterial behavioural outcomes suggest common behavioural parameters with other species under crucial circumstances. Among them are a response to environmental risk (fleeing, clustering, dormant, reproducing), convenient environment (approaching, increasing reproduction rate), acquiring resistance to hazardous agents, and invading compatible substrates. As if the Natural Kingdom would express a limited number of basic behavioural parameters in different forms under critical circumstances, according to the species' complexity.

On metaphorical grounds, and placing aside species pride, in these domains, human behaviour parallels that of bacteria with the addition of environmentally disturbing events with unknown long-term impacts on the species' future.

*

The fact that the two deepest branching in the eubacterial line of descent are both basically thermophilic and slowly evolving, strongly suggests that all eubacteria have ultimately arisen from a thermophilic ancestor.

(Achenbach-Richter 1987)

In this regard, perhaps a special mention should be given to those microorganisms labelled as "extremophiles" (tardigrades and bacteria) (see Figures 2.5 and 2.6), small but determined to contradict any reasonable and mundane thought about the conditions suitable for "life". They are dedicated to perpetrating themselves in extreme

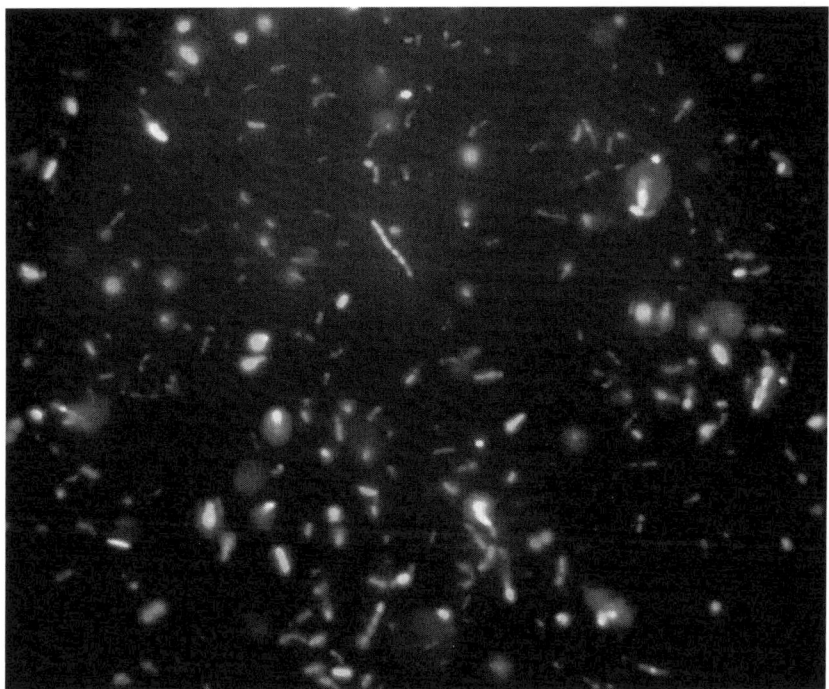

FIGURE 2.5 Extremophile bacteria was identified by NASA astrobiologist Dr. Richard Hoover after retrieval from ice dating back 32,000 years ago. (NASA/MSFC).[8]

FIGURE 2.6 Dr. Diane Nelson, a Tardigrade researcher who works in the Great Smoky Mountains National Park, took this 3-D image of a Tardigrade using a scanning microscope.[9]

environments in dark waters, without oxygen, with temperatures close to boiling or super cold, corrosive acidity levels, or nuclear reactors.

Early physical (thermal) and chemical conditions forced early life forms to adapt to survive them. Living forms able to survive extreme physical and chemical conditions are the *extremophiles.*

*

As stated by Douglas (2014) – although data are still fragmentary – several lines of evidence are consistent with the hypothesis that eukaryotes derive health benefits from symbiosis because their regulatory networks are structured to function in interactions with resident microbiota. Furthermore,

> ... success of the eukaryotes is founded on their acquisition of aerobic respiration and photosynthesis by symbiosis, with the transition of bacterial symbionts to organelles. The localization of electron transport chain in aerobic respiration and photosynthesis to the intracellular compartments of mitochondria and plastids, respectively, provided additional evolutionary opportunities for eukaryotes.

The recruitment of host genes by viruses is as common as the recruitment of viral genes. Pradeu (2011) considered that every organism is "mixed" heterogeneous. This requires a convergence of today's microbiology, immunology, ecology, and developmental biology that leads us to better understand the organism as the unity of such a plurality. In 2012, Gilbert et al. supported their concept that the discovery of symbiosis throughout the animal kingdom fundamentally transformed the classical concept of insular individuality into one with interactive relationships among species. This blurs the conceptual boundaries of an organism and obscures the notion of an essential identity. As Gilbert et al. (2015) commented, developmental symbiosis implies that organisms evolve through the interactions between the host and its persistent symbiotic microorganisms, probably involving plants and animals.

The multiple adaptive potential of microorganisms allowed the needed time to generate progressive, adaptable associations and build basic complex organisms. It also affects environmental conditions that allow multiple failure/success living trials of new species in interaction with the environment. This generated multiple living forms of variable adaptive degrees of freedom and adaptability to environmental cues (e.g., on comparative grounds, the reported differences in gut bacterial densities within species of ants depending on diet and habitat, as reported by Sanders et al., 2017). This sparked an interactive succession of biological events with different success rates, thus mutually conditioning their living probabilities and affecting the ecosystem.

Concerning the impact of viruses on the evolutive chain, according to Suttle (2007), the virosphere is probably inclusive of every environment on the Earth, from the atmosphere to the deep biosphere. As Koonin and Dolja (2013) stated, their general features represent one central point in the evolution process under which the history of life involving virus-host coevolution is one principal factor that defines the course of evolution of both cells and viruses. Viruses and/or virus-like elements are associated with all cellular life forms and are Earth's most abundant biological entities. This

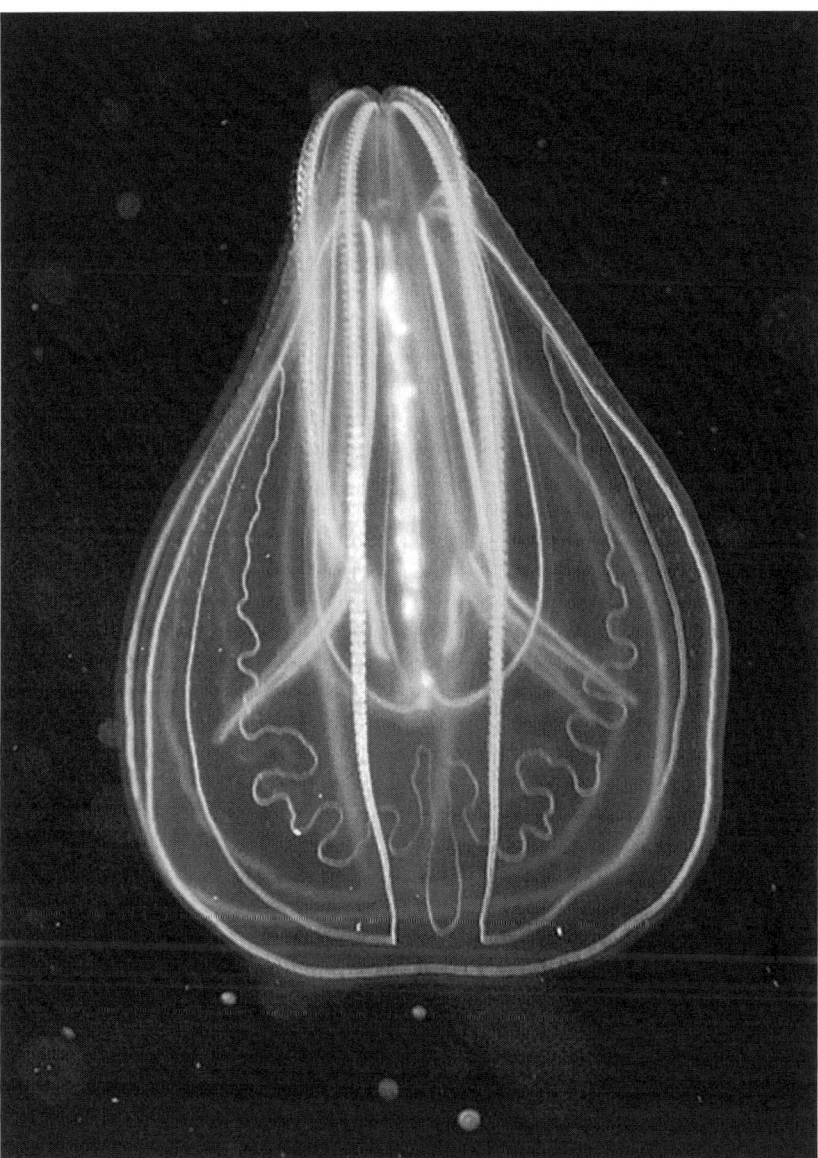

FIGURE 2.7 Comb jelly.[10]

coevolution likely started within primitive replicator systems and was essential for the major evolutionary transitions such as the emergence of DNA genomes, cellular organisation and later the eukaryotic cellular organisms.

> Although the virus world maintained its distinctness from the world of cellular organisms throughout the course of the evolution, the genetic exchange between the two parts of the biosphere always was extensive and highly consequential.
>
> *(Koonin and Dolja 2013)*

In closing this chapter, it seems adequate to bring up a statement that would imply the possibility of improving our understanding of the initial, ancient events leading to the formation of simple life forms based on controlled experimental conditions, a highly desirable prospect. The evolutive question of whether sponges or ctenophores (comb jellies) are the sister group to all other animals was approached by Schultz et al. (2023). This question implies different scenarios for the evolution of complex neural systems, as stated by the authors.

> Five major lineages arose early in animal evolution and survive to the present day: sponges (poriferans), ctenophores (comb jellies), placozoans (microscopic flat animals), cnidarians (such as anemones, jellyfishes and hydra) and bilaterians (such as chordates, molluscs, arthropods and diverse worms). Although morphological and phylogenomic studies consistently unite bilaterians, cnidarians, and placozoans into a monophyletic clade (Parahoxozoa) that excludes sponges and ctenophores, the relationship between sponges, ctenophores and Parahoxozoa remains controversial. There are two competing scenarios—the sponge-sister hypothesis and the ctenophore-sister hypothesis reflecting which lineage diverged first among animals.
>
> *(Schultz et al. 2023)*

Based on chromosome-scale gene linkage the authors concluded that comb jellies (see Figure 2.7) are the sister group to all other animals.

Notes

1 https://www.britannica.com/science/bacteria/Evolution-of-bacteria.
2 https://www.ncbi.nlm.nih.gov/books/NBK9841/.
3 https://earthathome.org/geologic-time-scale/.
4 https://commons.wikimedia.org/w/index.php?curid=123919392.
5 https://www.nature.com/articles/d41586-021-01316-0.
6 Sterols are a subgroup of steroids.
7 https://en.wikipedia.org/wiki/Horizontal_gene_transfer#/media/File:Tree_Of_Life_(with _horizontal_gene_transfer).svg.
8 https://www.nbcnews.com/id/wbna7019473.
9 https://www.nps.gov/grsm/learn/education/classrooms/lp-ncms-tardigrade-wrapup1. htm.
10 https://commons.wikimedia.org/wiki/File:Comb_jelly.jpg.

References

Achenbach-Richter, L., *et al.* "Were the Original Eubacteria Thermophiles?" *Systematic and Applied Microbiology*, vol. 9, no. 1–2, 1987, pp. 34–39, doi:10.1016/s0723-2020(87)80053-x.

Airapetian, V. S., *et al.* "Prebiotic Chemistry and Atmospheric Warming of Early Earth by an Active Young Sun." *Nature Geoscience*, vol. 9, no. 6, 2016, pp. 452–455, doi:10.1038/ngeo2719.

Andersson, G. E., *et al.* "On the Origin of Mitochondria: A Genomics Perspective." *Philosophical Transactions of the Royal Society of London. Series B: Biological Sciences*, vol. 358, no. 1429, 2003, pp. 165–179, doi:10.1098/rstb.2002.1193.

Biello, David. "Did Life's First Cells Evolve in Geothermal Pools?" *Scientific American*, 2012, www.scientificamerican.com/article/did-life-first-evolve-in-geothermal-pools/.

Brocks, Jochen J., *et al.* "The Rise of Algae in Cryogenian Oceans and the Emergence of Animals." *Nature*, vol. 548, no. 7669, 2017, pp. 578–581, doi:10.1038/nature23457.

Brocks, Jochen J., *et al.* "Lost World of Complex Life and the Late Rise of the Eukaryotic Crown." *Nature*, vol. 618, no. 7966, 2023, pp. 767–773, doi:10.1038/s41586-023-06170-w.

Brunet, Thibaut, and Nicole King. "The Origin of Animal Multicellularity and Cell Differentiation." *Developmental Cell*, vol. 43, no. 2, 2017, pp. 124–140, doi:10.1016/j.devcel.2017.09.016.

Butterfield, Nicholas J. "Early Evolution of the Eukaryota." *Palaeontology*, vol. 58, no. 1, 2014, pp. 5–17, doi:10.1111/pala.12139.

Cavalier-Smith, Thomas. "Origin of Mitochondria by Intracellular Enslavement of a Photosynthetic Purple Bacterium." *Proceedings of the Royal Society B: Biological Sciences*, vol. 273, no. 1596, 2006, pp. 1943–1952, doi:10.1098/rspb.2006.3531.

Dance, Amber. "The Mysterious Microbes That Gave Rise to Complex Life." *Nature*, vol. 593, no. 7859, 2021, pp. 328–330, www.nature.com/articles/d41586-021-01316-0, doi:10.1038/d41586-021-01316-0.

Dominguez-Bello, Maria Gloria, *et al.* "Role of the Microbiome in Human Development." *Gut*, vol. 68, no. 6, 2019, pp. 1108–1114, gut.bmj.com/content/68/6/1108, doi:10.1136/gutjnl-2018-317503.

Douglas, Angela E. "Symbiosis as a General Principle in Eukaryotic Evolution." *Cold Spring Harbor Perspectives in Biology*, vol. 6, no. 2, 2014, pp. a016113–a016113, cshperspectives.cshlp.org/content/6/2/a016113.full, doi:10.1101/cshperspect.a016113.

Dufourc, Erick J. "Sterols and Membrane Dynamics." *Journal of Chemical Biology*, vol. 1, no. 1–4, 2008, pp. 63–77, www.ncbi.nlm.nih.gov/pmc/articles/PMC2698314/, doi:10.1007/s12154-008-0010-6.

Edgell, David R., and W. Ford Doolittle. "Archaea and the Origin(S) of DNA Replication Proteins." *Cell*, vol. 89, no. 7, 1997, pp. 995–998, www.cell.com/action/showPdf?pii=S0092-8674%2800%2980285-8, doi:10.1016/s0092-8674(00)80285-80288.

Embley, T. Martin, and William Martin. "Eukaryotic Evolution, Changes and Challenges." *Nature*, vol. 440, no. 7084, 2006, pp. 623–630, www.nature.com/articles/nature04546, doi:10.1038/nature04546.

Forterre, Patrick, *et al.* "Luca : À la recherche du plus Proche Ancêtre Commun universel." *Médecine/Sciences*, vol. 21, no. 10, 2005, pp. 860–865, doi:10.1051/medsci/20052110860.

Gilbert, Scott F., *et al.* "A Symbiotic View of Life: We Have Never Been Individuals." *The Quarterly Review of Biology*, vol. 87, no. 4, 2012, pp. 325–341, doi:10.1086/668166.

Gilbert, Scott F., *et al.* "Eco-Evo-Devo: Developmental Symbiosis and Developmental Plasticity as Evolutionary Agents." *Nature Reviews Genetics*, vol. 16, no. 10, 2015, pp. 611–622, doi:10.1038/nrg3982.

Grosberg, Richard K., and Richard R. Strathmann. "The Evolution of Multicellularity: A Minor Major Transition?" *Annual Review of Ecology, Evolution, and Systematics*, vol. 38, no. 1, 2007, pp. 621–654. doi:10.1146/annurev.ecolsys.36.102403.114735.

Kappler, Andreas, *et al.* "Deposition of Banded Iron Formations by Anoxygenic Phototrophic Fe(II)-Oxidizing Bacteria." *Geology*, vol. 33, no. 11, 2005, p. 865, doi:10.1130/g21658.1.

Keeling, Patrick J., and Jeffrey D. Palmer. "Horizontal Gene Transfer in Eukaryotic Evolution." *Nature Reviews Genetics*, vol. 9, no. 8, 2008, pp. 605–618, www.nature.com/articles/nrg2386#f1, doi:10.1038/nrg2386.

Keeling, Patrick J. "The Number, Speed, and Impact of Plastid Endosymbioses in Eukaryotic Evolution." *Annual Review of Plant Biology*, vol. 64, no. 1, 2013, pp. 583–607, doi:10.1146/annurev-arplant-050312-120144.

Koonin, Eugene V. "On the Origin of Cells and Viruses." *Annals of the New York Academy of Sciences*, vol. 1178, no. 1, 2009, pp. 47–64, doi:10.1111/j.1749-6632.2009.04992.x.

Koonin, Eugene V., and Valerian V. Dolja. "A Virocentric Perspective on the Evolution of Life." *Current Opinion in Virology*, vol. 3, no. 5, 2013, pp. 546–557, doi:10.1016/j.coviro.2013.06.008.

Ley, Ruth E., *et al.* "Ecological and Evolutionary Forces Shaping Microbial Diversity in the Human Intestine." *Cell*, vol. 124, no. 4, 2006, pp. 837–848, doi:10.1016/j.cell.2006.02.017.

Mansy, Sheref S., *et al.* "Template-Directed Synthesis of a Genetic Polymer in a Model Protocell." *Nature*, vol. 454, no. 7200, 2008, pp. 122–125, www.ncbi.nlm.nih.gov/pmc/articles/PMC2743009/, doi:10.1038/nature07018.

Martin, William, and Michael J. Russell. "On the Origins of Cells: A Hypothesis for the Evolutionary Transitions from Abiotic Geochemistry to Chemoautotrophic Prokaryotes, and from Prokaryotes to Nucleated Cells." *Philosophical Transactions of the Royal Society of London. Series B: Biological Sciences*, vol. 358, no. 1429, 2003, pp. 59–85, doi:10.1098/rstb.2002.1183.

Mojzsis, S., *et al.* "Evidence for Life on Earth Before 3,800 Million Years Ago." *Nature*, vol. 384, 1996, pp. 55–59, doi:10.1038/384055a0.

Porter, Susannah M. "Insights into Eukaryogenesis from the Fossil Record." *Interface Focus*, vol. 10, no. 4, 2020, p. 20190105, doi:10.1098/rsfs.2019.0105.

Pradeu, Thomas. "A Mixed Self: The Role of Symbiosis in Development." *Biological Theory*, vol. 6, no. 1, 2011, pp. 80–88, doi:10.1007/s13752-011-0011-5.

Sanders, Jon G., *et al.* "Dramatic Differences in Gut Bacterial Densities Correlate with Diet and Habitat in Rainforest Ants." *Integrative and Comparative Biology*, vol. 57, no. 4, 2017, pp. 705–722, doi:10.1093/icb/icx088.

Schultz, Darrin T., *et al.* "Ancient Gene Linkages Support Ctenophores as Sister to Other Animals." *Nature*, vol. 618, 2023, pp. 110–117, www.nature.com/articles/s41586-023-05936-6, doi:10.1038/s41586-023-05936-6.

Snel, Berend, *et al.* "Genomes in Flux: The Evolution of Archaeal and Proteobacterial Gene Content." *Genome Research*, vol. 12, no. 1, 2001, pp. 17–25, doi:10.1101/gr.176501.

Suttle, Curtis A. "Marine Viruses – Major Players in the Global Ecosystem." *Nature Reviews Microbiology*, vol. 5, no. 10, 2007, pp. 801–812, doi:10.1038/nrmicro1750.

Thompson, Luke R., *et al.* "A Communal Catalogue Reveals Earth's Multiscale Microbial Diversity." *Nature*, vol. 551, no. 7681, 2017, pp. 457–463, doi:10.1038/nature24621.

Whitman, W. B., *et al.* "Prokaryotes: The Unseen Majority." *Proceedings of the National Academy of Sciences*, vol. 95, no. 12, 1998, pp. 6578–6583, www.cen.ulaval.ca/merge/pdf/Whitman1998.pdf, doi:10.1073/pnas.95.12.6578.

Woese, C. R., and G. E. Fox. "Phylogenetic Structure of the Prokaryotic Domain: The Primary Kingdoms." *Proceedings of the National Academy of Sciences*, vol. 74, no. 11, 1977, pp. 5088–5090, doi:10.1073/pnas.74.11.5088.

Woese, C. R. "Bacterial Evolution." *Microbiological Reviews*, vol. 51, no. 2, 1987, pp. 221–271, doi:10.1128/mmbr.51.2.221-271.1987.

Woese, C. R., *et al.* "Towards a Natural System of Organisms: Proposal for the Domains Archaea, Bacteria, and Eucarya." *Proceedings of the National Academy of Sciences of the United States of America*, vol. 87, 12, 1990, pp. 4576–4579, doi:10.1073/pnas.87.12.4576, PMID: 2112744, PMCID: PMC54159.

Woese, C. R. "The Universal Ancestor." *Proceedings of the National Academy of Sciences of the United States of America*, vol. 95, no. 12, 1998), pp. 6854–6859, doi:10.1073/pnas.95.12.6854.

3

THE DYNAMIC OF CHANGE FROM THE HUMAN PERSPECTIVE

*

For illustration purposes, let us suppose that our hypothetical book about the planet's natural history – to which we wish to contribute as a species – had up to 100 pages. Suppose also that each page is equivalent to the passage of about 50 million years. The book total would then count as 5 billion years, a figure close to our planet's supposed age. It is estimated that life in its most primary form – a particulate, organised entity with the ability to self-replicate – would have existed for about 3.5 billion years. In other words, of the book's 100 pages, about 35 would have elapsed before being able to read about an elementary form of life. Multicellular organisms would have appeared around a billion years ago, perhaps by page 80 of our hypothetical book. The age of the great reptiles died out about 60 million years ago. In this book, such a catastrophic event only occurred next to the last page. Then, mammals began to propagate and became the planet's dominant species. Primates began to leave their footprint about 30 million years ago; the last half page of our hypothetical book would represent that. If each page had 50 lines, the primates' story would take up the last 30 lines of this book – a million years per line of text.

The hominoids would have separated from the great apes about 6 million years ago – six lines before the end of the book. *Homo erectus* would have walked the African savannas no more than 2 million years ago – two lines before the end of the book. If, again, we agree for simplicity that 100 characters with spaces make up a line of ordinary text, each character would be equivalent to 10,000 years. If we assign *Homo sapiens* no more than 200,000 years, that would be a fifth of the last line of text – exactly 20 characters. The written history of *H. sapiens* is about 6,000 years old; that is less than an entire letter. In that relatively brief history, we have created knowledge and beauty, placed the planet and our survival at risk, and ruined the lives of billions of people.

Let us place this evolution in another imaginary context. Stringer (2005) uses the calendar metaphor to illustrate the recent emergence of humans. If the estimated 4.56

DOI: 10.4324/9781032698380-3

billion years of the planet's existence were compressed into a calendar year, each day would represent the passage of 12.5 million years. Each minute would represent 8,500 years. In that calendar, by noon on the last day of December, the hominoids and the chimpanzee would have diverged from their common trunk. By mid-afternoon, the genus *Homo* would have diverged from the australopithecines. With 20 minutes to spare before the end of the year, the first modern humans would have appeared. A minute before midnight, the first inhabitants of the American continent would appear. Our modern time would be represented by one second before the last chime.

Let us try a third staging of events, focusing on the *Homo sapiens*. Our species is credited with an approximate existence of about 200,000 years since modern times. The written history of humans would have appeared approximately 6,000 years ago. The steam machine was invented in the 18th century, just over 200 years ago. The technological revolution and the investigation of outer space have been active for almost 80 years. Exploratory probes have successfully reached Mars, Saturn, and Venus. Nanotechnology and genetic engineering have about 50 to 70 years of development. If the unrestrained acceleration of technological development of *sapiens* does not lead us to crash into the proverbial wall, what possible future awaits us?

How far is it possible to speculate on *Homo sapiens*' future? There are at least two main aspects to elaborate on – productive and destructive. Is it possible that the contributions of the first will cancel out the second? Suppose one observes the record of surviving and extinct species, the variety of adaptation resources, their endless numbers, their incredible adaptation to different ecological niches, and the persistent, primaeval survival of microbiological life. One cannot avoid asking at least a couple of additional and uncomfortable questions, which the author will avoid.

*

> If we consider ourselves to be a composite of microbial and human species, our genetic landscape a summation of the genes embedded in our human genome and microbiome, and our metabolic features a coalescence of human and microbial traits, the self-portrait that emerges is one of a "human supraorganism".
>
> *(Turnbaugh et al. 2007)*

As will be expanded in the following chapters, humans are built by a historical succession of genetic codes with their combined expression modified through aeons, a series of environmental interactions designed to take the best possible options to survive, including exerting dominance to assure survival. Human survival operates within this narrow bandwidth of biological heritage. Blind evolution implies the possibility of eventually perfecting this interaction and adaptive process, with winners achieving the best survival probability.

The human quota of decisions is trapped within a band of possible behaviours whose profiles express this binding to exert competing survival options. Freedom is represented by this limited quota of possible decisions bound to individual and collective historical experience. Technological evolution is nothing less than an instrument to accomplish improved survival and an attempt to perfect and spread the dominant macro-species and expand human knowledge beyond its senses. Yet, this spread will carry primaeval, fundamental, living forms that conform to the Natural Kingdom's genesis. There is no genius brain behind this evolution or survival purpose, but a succession of codes represents the dawn of life on this planet, where different species performed the blind role of transient carriers of essential, replicating life forms. So far, humans provide the most imaginative and creative species that could carry our "simple" commensals and genetic codes to distant times and territories.

*

Besides commensal forms required for complex living systems, cellular respiratory chains are intracellularly localised in eukaryotic cells, in mitochondria. These organelles (endosymbionts) originated from the permanent enslavement of purple non-Sulphur bacteria (Cavalier-Smith, 2006).

It could be considered that the ultimate concept of dominance is survival, and so far, as an order, archaeons, bacteria, and primitive eukaryote forms have outlived all complex organisms that evolved from them. Living creatures originated following the genesis of simple, primal life forms colonising Earthly primaeval substrates and surviving unimaginable demanding ecological events. Nature events provided grounds to generate continuously adapting and evolving life forms – the emergence of microspecies – able to adapt, survive, and multiply to various substrate and climate conditions. With increasing complexities, macrospecies emerged following a series of survival failures due to geoclimatic events or exhaustion/inadequate adaptive mechanisms; this was a repeated event that evolved through a sequence of geological eras. Before each of them, creatures adapted and roamed on a planet that recreated terrain and climate conditions.

Wilson (2012) asserts that termites would have arisen about 220 million years ago in the mid-Jurassic, ants about 150 million years ago in the late Jurassic to early Cretaceous, and honeybees and bumble bees 70–80 million years ago during the late Cretaceous. Interestingly, the rise of the Hymenoptera and Isoptera (termites) and assorted insects in some of the other orders was accompanied by the development of some form of social behaviour.

Should we consider social behaviour and ecological adaptive processes, an interactive domain of mutual influences (whether commensal or predatory), we should also include plant-plant interactions, as was pointed out by Brooker (2010), who places these interactions in a broader context,

> Given the above considerations, it perhaps would be convenient to view these multiple relationships involving survival-bound micro- and macro-organisms, whether of animal or plant origin, as opportunistically generating an integrated equilibrium or mutual adaptation. An equilibrium that the human species took good care to disarray and condemn to a condition of fragility.

Additionally, as stated by Koskella and Bergelson (2020),

> The importance of species interactions in shaping organismal diversity, species ranges, community structure and ecosystem function is a long-standing focus in

ecology and evolutionary biology. This is due not only to the known impact of species interactions on the evolutionary potential of populations and communities, but also to the possibility that these interactions result in novel functions.

*

The above comments bring up the concept of the Anthropocene, the era dominated by human actions affecting all levels of natural life. Let us place it within the general view of geobiological developments.

Considering the evolution of life, the expression of our original genome evolved to adapt to harsh, changing conditions. Hence, some social and environmental changes of our modern era have occurred too rapidly on biological terms to adequately adapt our genomic inertias that originated in past horizons. This mismatch between ancient genomic instructions and current socioecological demands implies potential consequences in the physical and behavioural domains. Our adaptation capacity is fenced by genetic and physiological constraints, with the addition that most of them have been moulded to ancient environmental demands exerted during aeons. Thus, besides human limited cognitive and emotional plasticity, its physiological stimulus-response *black box* was shaped to confront ancient conditions and demands from aeons past that we now implement to cope with modern demands. On metaphorical grounds, we implement and try to adapt an old racing car to perform on a modern Indianapolis racing track (or similar ones) by adjusting some construction pieces without previous collective adaptation and training. This is generally announcing a fractured relationship between our social collective behavioural habits, mental resources, and scientific-technological progress.

FIGURE 3.1 Ediacaran fossil from the Nilpena Ediacara National Park in Australia. Scott Evans photography. *Ediacara* would be the oldest community of complex, macroscopic fossils.[2]

According to statements on what would have been the first mass extinction, the relationship of a geological era with species extinction rates would seem to remain under continuous revision. As reported by Evans et al. (2022) and commented by Krish (2022)[1], a global drop in oxygen levels about 550 million years ago would have led to Earth's first known mass extinction (involving creatures like the one shown in Figure 3.1), leading to 80% of species eradication linked to climate change, leaving almost no traces in the fossil record until recent reports. This early mass life extinction would later be followed by several others, including the Ordovician-Silurian (named after an ancient Celtic tribe, the *Ordovices*, and following a Welsh tribe, the *Silures,* where first rocks were described from this era) and the Devonian extinctions (a name given due to reddish rocks that were first studied in Devon, England) (from 440 million and 365 million years ago, respectively) (see Figure 3.1 for geologic time scales), killing many marine organisms. Later, the Permian-Triassic and Triassic-Jurassic extinctions – approximately 250 million and 210 million years ago, respectively – affected ocean vertebrates and land animals. More recently, about 66 million years ago, at the end of the Cretaceous period, extinction wiped out approximately 75% of plants and animals, including non-avian dinosaurs. Thus, the emergence and survival of our species represent an event within a process of periodic massive extinctions, to which our mindful species will have to find solutions for its long-term survival, starting by avoiding self-destruction.

*

The Concept of Anthropocene

Though the concept of Anthropocene still sparks debates on its geological time frame and conceptual identity – as mentioned by Issberner and Léna (2018) –, according to UNESCO, the term Anthropocene was coined to consider the impact of the accelerated accumulation of greenhouse gases on climate and biodiversity, and the irreversible damage caused by the over-consumption of natural resources.

According to The UNESCO Courier (2018),

> … you are living in the Phanerozoic eon, the Cenozoic era, the Quaternary period and the Holocene Epoch. These are all subdivisions of our planet's geological time scale to which a new bar will probably soon be added, the Anthropocene.

However, do we need to turn this into a new geological epoch? There is no question that,

> The competitive globalized system tries to find supplies at the least cost, encouraging extractivism in many countries, and land-grabbing in others… This is a sign of the weak governance of the climate question which, in the absence of an institution mandated to carry it out, is incapable of prevailing over the economic interests of countries and enterprises.[3]

As denounced by Lena and Issberner (2018),[4]

Submerged under contradictions, dilemmas and ignorance, the extremely serious environmental issues of the Anthropocene are not getting the required level of priority in national social agendas.

It all began following an untestable origin of simple, primal life forms colonising Earthly substrates and surviving unimaginable, demanding ecological events. At some point, nature events provided grounds to start continuously adapting and evolving more complex life forms (thus, able to adapt, survive, and multiply their living units) to various substrate and climate conditions. They were all colonised by or carried commensal forms of bacteria that would *dispatch* or disregard their carrier when its physiological system would have decayed and ceased living.

Survival price was high, for uncounted terrestrial, water, and aerial species were occasionally banned. However, the cost allowed for rebuilding each time fauna and ecosystems adapted under different climate and ground parameters. According to Alegado et al. (2012), sulfonolipids – a class of sphingolipids featuring a sulfonate group – produced by bacteria would have been involved in triggering eukaryotic morphogenesis and thus involved in the evolution of multicellular organisms. Natural history generated a continuum of new species that lived and failed to survive at different rates. Animal macrospecies shined at turns and found their lives cancelled in a continuous interaction with the ecosystem and their adaptive capacity span. No one had a firm living contract in a changing world plagued with competitors thriving to adapt and survive. Species flourished and disappeared to join the remains of uncountable ancestors or unrelated partners. No life insurance was available, only a limited number of species resources to transiently adapt and prevail in a competitive world.

However, were all natural history specimens' survival efforts and experiences useless? And thus, drowned in a blind cesspool? Moreover, closer to our interests, have our species been built with the interactive summary of all that, yet almost unaccounted, precise history? Is our history as *Homo sapiens* another *brick in the wall* of Natural History? Even though the answer would be in the affirmative, we humans appear to have not yet assumed the implications of such conceptually appalling history projected into our species construction, for it may preannounce the probable limits and conditions of our freedom to evolve any further as a biological species, perhaps into forms and behaviours with which we would not recognise as ourselves.

All species carry a basic set of building elements that give way to variable survival strategies. They do not resemble drops of water that passively adopt the shape of its container, and yet their adaptive survival strategies depend on their own physiological constructs and the limits imposed by the external demands ("the container"). More precisely, it is the variety of species, their adaptive spread and behavioural innovation that, in the long term, somehow provide some assurance of the continuation of life in a context of changing ecosystems and collective interactions.

What is the impact of humans on this *survival equation*? On a global view of human actions, are we providing an element that potentiates an ecological and living species equilibrium and survival? Or are we blindfolded to our impact on our universal survival conditions? In such a perspective of shared, interactive species survival, do our drives and actions belong to *the wrong species* (Colombo 2007) since we act following our selfish goals and profit interests irrespective of its environmental and ecological consequences?

> ... the notion that rapid, and marked, transformations in human lifestyles are not only affecting the health of the biosphere, but possibly our own health as a result of changes in our microbial ecology.
>
> *(Turnbaugh et al. 2007)*

In the following chapters, it will be attempted to expose some convivial and evolutive profiles that influence social construction and provide a different look to our ill-based human pride as members of the Natural Kingdom as we enter the concept of host-gut microbiome interactions.

Notes

1 https://www.livescience.com/1st-mass-extinction-oxygen-drop.
2 https://www.livescience.com/1st-mass-extinction-oxygen-drop.
3 https://courier.unesco.org/en/articles/anthropocene-vital-challenges-scientific-debate.
4 https://en.unesco.org/courier/2018-2/anthropocene-vital-challenges-scientific-debate.

References

Alegado, Rosanna A., *et al.* "A Bacterial Sulfonolipid Triggers Multicellular Development in the Closest Living Relatives of Animals." *eLife*, vol. 1, 2012, doi:10.7554/elife.00013.

Brooker, Rob. "Plant Communities, Plant–Plant Interactions, and Climate Change." In *Positive Plant Interactions and Community Dynamics*, edited by Francisco Punaire. London, New York: Taylor & Francis, 2010, pp. 99–123, doi:10.1201/9781439824955-c6.

Cavalier-Smith, Thomas. "Origin of Mitochondria by Intracellular Enslavement of a Photosynthetic Purple Bacterium." *Proceedings of the Royal Society B: Biological Sciences*, vol. 273, no. 1596, 2006, pp. 1943–1952, doi:10.1098/rspb.2006.3531.

Colombo, Jorge A. *Pobreza y Desarrollo Infantil. Una Contribucion Multidisciplinaria*. Buenos Aires: Ediciones Paidós, 2007, pp. 97–113.

Evans, Scott D., *et al.* "Environmental Drivers of the First Major Animal Extinction across the Ediacaran White Sea-Nama Transition." *Proceedings of the National Academy of Sciences*, vol. 119, no. 46, 2022, doi:10.1073/pnas.2207475119.

Issberner, Liz-Rejane, and Philippe Léna. "Anthropocene: The Vital Challenges of a Scientific Debate." *The UNESCO Courier*, 15 May 2018, https://courier.unesco.org/en/articles/anthropocene-vital-challenges-scientific-debate.

Koskella, Britt, and Joy Bergelson. "The Study of Host–Microbiome (Co)Evolution across Levels of Selection." *Philosophical Transactions of the Royal Society B: Biological Sciences*, vol. 375, no. 1808, 2020, p. 20190604, doi:10.1098/rstb.2019.0604.

Krisch, Joshua A. "Scientists Just Found a Hidden 6th Mass Extinction in Earth's Ancient Past." *LiveScience*, 2022, www.livescience.com/1st-mass-extinction-oxygen-drop.

Stringer, Chad. *The Complete World of Human Evolution*. New York: Thames & Hudson, Inc, 2005.

The UNESCO Courier, "Waiting for the Heroes to Come." April–June 2018.

Turnbaugh, Peter J., *et al.* "The Human Microbiome Project." *Nature*, vol. 449, no. 7164, 2007, pp. 804–810, www.ncbi.nlm.nih.gov/pmc/articles/PMC3709439/, doi:10.1038/nature06244.

Wilson, Edward O. *The Social Conquest of Earth*. New York: Liveright Pub. Corp, 2012.

4

ON THE HOLOBIONT AND HOLOGENOME

Preliminary Comments

> Animals comprise a relatively recently derived clade and descend from an ancestral lineage that underwent extensive loss of genes underlying metabolic capabilities, including pathways required for biosynthesis of molecules essential to life. This limited metabolic toolkit is relatively constant across animal species, reflecting loss in shared ancestors followed by low rates of acquisition of novel genes through horizontal transfer. In contrast, species of bacteria and other microorganisms collectively possess a vast array of pathways for consuming and creating organic molecules, along with sophisticated molecular machines for delivering their gene products to influence nearby organisms.
>
> *(Moran et al. 2019)*

The basic motion leverage for evolution in the Natural Kingdom is based on a random, opportunistic, blind usage of conditions that allow survival means, as characterised by Dawkins in *The Selfish Gene* (1989). Though a survival goal provides different innate behavioural, strategic profiles according to the considered animal or vegetal species, such construction recognises aeons of trial and error – of survival success and failure to do so – on this interaction between individuals and the ecosystem. This survival behaviour originally started as the ability of *simpler* organisms (Archean, Prokaryotes, and Eukaryotes) to survive and reproduce in ecological environments of different physical and chemical profiles. Millions of years later, with the emergence of progressively complex macro-organisms, a new organic universe reset survival opportunities and strategies for these early living, morphologically basic living forms. After successful and negative survival trials practised during aeons, eukaryotic macro-organisms evolved based on association or commensalism with Archean and prokaryotes or becoming endosymbionts of plant and animal cells. This allowed primal living forms to endure in different media coupled with progressively complex carriers prepared to survive with a limited survival-set under different ecosystems. No living creature is devoid of interaction with these opportunistic survivors stemming from remote times in natural life history.

DOI: 10.4324/9781032698380-4

In most cases, this association developed a condition of mutual providers of molecules or functions (as the mitochondrion in the intracellular respiratory chain). These simple, unicellular ancient inhabitants of the Natural Kingdom became a needed component for the evolution of complex macro-organisms. This process could be considered an additional survival strategy that proved successful for millions of microbes installed in the core of complex, multicellular individuals.

Hence, in evolutionary domains, these diverse forms (symbionts) sharing a host physiological environment and adapting to it became necessary for individual survival. It can be considered a successful means of colonising congenial macro-organisms of different complexity adapted to survive in most ecosystems. A sort of useful carrier that provided a new ecological platform for their own survival and expansion in the Natural Kingdom which provided a new perspective to the concept of species evolution. As expressed by Luckey (1972),

> Animal evolution provided increased numbers, kinds, and efficiencies of microbic incubators.

A further statement by Luckey (1972) provides a perspective for the evolutive gain on this successful, evolutive strategy of simple cell forms, i.e., to *associate with* any emerging complex organism,

> … one attribute of life is that each ecologic niche fills with the most living matter possible, within the idiosyncrasies of certain vertebrates. Planet earth probably became filled with microbes during its second and third billion years of existence. Metazoans developed in this microbic milieu during the fourth billion years of earth life. A mutualistic symbiosis developed. Metazoans, particularly vertebrates, provided accommodation for certain microbic species on their external surfaces and in their alimentary tracts. Those microbic species that could successfully occupy the alimentary tract found a new kind of ecologic niche.

*

Holobiont and Hologenome

Though the microbiome interacts with developmental, physiological, and phenotypic domains, its contribution to host adaptation remains unsettled. As stated by Huitzil et al. (2018), there is undeniable evidence showing that bacteria have strongly influenced multicellular organisms' evolution and biological functions. Under this premise, *holobiont* concept was proposed, i.e., the host plus the microbiota, and its role in several physiological and behavioural domains was signalled. This concept should include ecological and evolutionary domains that shape gut microbial density, as posed by Ley et al. (2006).

Rosenberg and Zilber-Rosenberg (2018) stated that,

> Consideration of the holobiont with its hologenome as an independent level of selection in evolution has led to a better understanding of underappreciated modes of genetic variation and evolution. The hologenome is comprised of two

complimentary parts: host and microbiome genomes. Changes in either genome can result in variations that can be selected for or against. The host genome is highly conserved, and genetic changes within it occur slowly, whereas the microbiome genome is dynamic and can change rapidly in response to the environment by increasing or reducing particular microbes, by acquisition of novel microbes, by horizontal gene transfer, and by mutation.

As suggested by Henry et al. (2021), the *hologenome* theory – as initially proposed and discussed in Zilber-Rosenberg and Rosenberg (2008) and Rosenberg and Zilber-Rosenberg (2018) – integrates microbial genetic variation into the host evolutionary process. The hologenome is an evolutionary unit combining the eukaryotic host and associated microbes. Under this concept, evolution operates on this single unit because eukaryotic hosts are never isolated from microbes in the natural world (Rosenberg and Zilber-Rosenberg 2018), and thus, the evolutionary fate of both hosts and microbes is intertwined. The authors also considered that genetic variation in the hologenome can be brought about by changes in the host genome and in the microbiome genome. Since the microbiome genome can adjust to environmental dynamics more rapidly and by more processes than the host genome, it can play a fundamental role in the adaptation and evolution of the holobiont. According to these authors, molecular studies have demonstrated that genetic variation and the evolution of holobionts involve the acquisition of novel microbes and microbial genes into host chromosomes.

As Zilber-Rosenberg and Rosenberg (2008) suggested, eukaryotes probably arose from prokaryotes and remained associated as a *holobiont*, considered to be the animal or plant with all its associated microorganisms as a unit of selection in evolution. These authors propose the concept of *hologenome*, defined as the sum of the genetic information of the host and its microbiota. According to these authors, the theory is based on four generalisations:

1. All animals and plants establish symbiotic relationships with microorganisms.
2. Symbiotic microorganisms are transmitted between generations.
3. The association between host and symbionts, considered a unit of selection in evolution, affects the fitness of the *holobiont* within its environment.
4. Variation in the *hologenome* can be brought about by changes in the genomes of either the host or the microbiota.

As reported by Wong et al. (2015),

> An often-neglected feature is the tension and disparity in biological scales between the host and its symbionts: the animal cells dominate the biomass, are typically isogenic (but phenotypically differentiated) and many have slow turnover; whilst the microbial cells are of lower biomass, up to orders of magnitude greater in numbers, genetically diverse and have faster turnover (Whitman et al., 1998; Ley et al., 2006; Grice et al., 2008).

The size of bacterial presence in human physiology (*the forgotten organ* of O'Hara and Shanahan, 2006), in terms of additional genetic material, was approximated by Zilber-

Rosenberg and Rosenberg (2008) to be of the order of 250,000 unique bacterial genes, compared to the host genome that would contain about 20,500 genes (She et al. 2004), thus providing to the hologenome a considerable increase in the potential of change.

This genomic combination would provide an extended environmental capacity to compete, expressing a metabolic activity equal to a *virtual organ within an organ,* according to Bocci (1992).

Let us remember that according to Rosenberg and Zilber-Rosenberg (2018),

> … microbes were on this planet for 2.1 billion years before there were any animals or plants. During this time, they evolved enormous biochemical diversity and split into two domains, Bacteria and Archaea. The first eukaryote was probably formed by the acquisition of bacteria to eventually form mitochondria and chloroplasts and possibly by the uptake of an Archaea by Bacteria to form the nucleus. Uptake of microbes into multicellular organisms continued to provide genetic variation for holobionts throughout evolution. Many of the beneficial interactive fitness traits of holobionts discussed above fit into this category.

It seems worthwhile to consider further the holobiome concept from two different viewpoints.

One, as described by Rosenberg and Zilber-Rosenberg (2018), is its role of providing a further adaptive range to the host. This is based on the dynamic characteristics of the microbiome genome that provides a domain able to change rapidly in response to the environment by increasing or reducing its numbers or acquiring novel microbes by horizontal gene transfer and mutation. On the contrary, the host genome is highly conserved, and genetic changes occur comparatively slowly. According to the authors, these combined profiles would allow holobionts to adapt and survive under changing environmental conditions, providing the time necessary for the host genome to adapt and evolve. This concept of the holobiont with its hologenome as an independent level of selection in evolution proposed a new understanding of genetic variation and evolution modes.

The second implicit domain of this holistic or unitary concept of host and microbiome consolidates a process that proceeded during aeons on all living species. That is, where the initial, comparatively simpler, living forms (the microbiome) spread their presence and expand their survival conditions by hijacking different micro- and macro-species (vegetal and animal). This is a sort of indirect expansion of its domains since their early appearance on a primal Earth. This adds complexity to our biological concept of living species' mutual dependence interactions, including humans. From an instrumental conceptualisation of macro-species carriers for these primal living forms emerge the construction that the drivers towards progressive more complex organisms were shared by the ubiquitous, mutagenic specimens of Archean, prokaryotes, and eukaryotes invading or hijacking more complex organic carriers roaming on different ecological niches, thus generating the holobionts.

According to Rosenberg and Zilber-Rosenberg (2018), considerable evidence supports the hypothesis that holobionts, with their hologenomes, can be considered levels of selection in evolution. According to these authors (cf. Rosenberg and Zilber-

Rosenberg, 2013, in Münger et al., 2018), the holobiont concept suggests that (i) all macro-organisms harbour microorganisms, serving the latter as nutrient-rich environments where they thrive; (ii) the fitness of the holobiont and its symbionts is interdependent; (iii) the hologenome can change due to variations in either the host's genome or the microbiome; and (iv) modifications are transmissible across generations and may thus influence the holobiont's evolution. According to Risely (2020), the term *microbiome* would have evolved to encompass variable definitions, often identifying critical microbes concerning their spatial distribution, temporal stability or ecological influence, and their contribution to host function and fitness.

As it was put forward earlier by McFall-Ngaia et al. (2013),

> … the current-day relationships of protists with bacteria, from predation to obligate and beneficial symbiosis, were likely already operating when animals first appeared. Attention to this ancient repertoire of eukaryote–bacterial interactions can provide important insights into larger questions in metazoan evolution, from the origins of complex multicellularity to the drivers of morphological complexity itself.

As Shapira (2016) also stated,

> It is well accepted that host–symbiont coevolution is responsible for fundamental aspects of biology. However, the emerging importance of plant- and animal associated microbiotas to their hosts suggests a scale of coevolutionary interactions many-fold greater than previously considered.

Douglas (2014) considered that for most associations, the effects of symbiosis can be attributed to providing a source of novel capabilities (traits carried by the microbial domain) and the fitness impact due to microbial modulation of growth rates, immune function, nutrient allocation, and behaviour.

Sarkar et al. (2020) proposed that multicellular hosts microbial life and the relationships between microorganisms and host lineages appear stable over millions of years of host evolution.

*

As commented, Foster et al. (2017) argued on this unitary evolving concept of the holobiont and further questioned (Sharp and Foster 2022) if mutual benefits of host-microbe relationships can explain cooperative evolution. The concept or metaphor of a *superorganism* composed of the host and the microbiome acting in concert with mutual convenience has been challenged by Foster et al. (2017). These authors support the concept that the host and each microbial strain are distinct entities with potentially divergent selective pressures and, hence, are single evolutionary units acting with a common interest. The authors further contrast the microbiome – an ecosystem held on a leash by the host – to three alternatives: a host-control model, a symbiont-control model, and an open-ecosystem model. From a mutual survival viewpoint, though, there ought to appear a point of mutual *system's* equilibrium based on feedback signals (as in any complex organism or the ecosystem itself); otherwise, an untenable, unstable condition would result in an imbalance and mutual survival challenge (i.e., the needed *leash* suggested by the authors?). The components would operate as in a "holobiont" at such a feedback equilibrium point. The mentioned authors further consider that their model predicts mutual benefits are insufficient to drive cooperation in systems like the human microbiome because of competition between symbionts. Should hosts exert control over symbionts, cooperation could emerge, so long as constraints limit symbiont counter-evolution. In this domain, regarding the holobiont, Vliet and Doebeli (2019) proposed that the concept that hosts and their microbiome form an integrated evolutionary entity, on which selection can potentially act directly, at best, is controversial based on whether the association between hosts and their microbiomes is strong enough to allow for selection at the holobiont level as a single unit.

*

According to Moeller and Sanders (2020), variation in the gut microbiota within mammalian species can profoundly affect host phenotypes. Its association would have a series of consequences, such as promoting the diversification of mammalian species by enabling a series of functional events. It would additionally include further functional impacts, such as dietary transitions onto difficult-to-digest carbon sources and toxic food items, shaping the evolution of adaptive phenotypic plasticity in mammalian species through the amplification of signals from the external environment and postnatal developmental processes. Additionally, it would generate selection for host mechanisms, including innate and adaptive immune domains, to control the gut microbiota for the benefit of host fitness. These effects would depend on the specific gut microbiotes within host species lineages, which could vary across mammalian phylogeny and could harbour compositionally distinct gut microbiotes.

> The human holobiont is progressively being understood, as the collective microbiome and host functions are better characterised in health and disease, and as we assess both correlation and causal relationships.
>
> *(Dominguez-Bello et al. 2019)*

However, as Zilber-Rosenberg and Rosenberg (2008) stated, while individual organisms evolve by selection of random variants, the proposed holobiont can evolve by adaptive processes. This implies a continuous interaction of variable environmental conditions and individual internal changes with development and ageing factors, affecting the interactive web's function. As Henry et al. (2021) expressed, while microbiomes have substantial phenotypic effects on their hosts, these effects strongly depend on the ecological (physiological) context, the additional compartment of the natural web. The latter involves hormones, neurotransmitters, receptors, humoral clearance, diet restrictions, immunological conditions, physiological recovery processes, and microbiome composition. Thus, the equation of biological interactions must add a factor (development, ageing) that alters the predictability of the interactive web mentioned above. Hence, according to the hologenome concept, variation, an important factor in evolution, can be brought about by modification in either the host or the microbiota genomes or intervening changes in the larger web of interactions.

As stated by Wong et al. (2015), based on the holobiont concept that anticipates trade-offs between host and gut microbiote, dietary choices and feeding habits are affected to prioritise specific host-gut microbiome interactions. As described in experimental designs (see also Slyepchenko et al. 2015; Burokas et al. 2015), the latter can be affected by probiotic supplements, whose impact on cognition was shown in ageing experimental rats by O'Hagan et al. (2017), and on lead-induced memory deficits in female rats prenatally exposed to probiotic by Zhang et al. (2023).

In this interactive, evolutive domain, major transitions were summarised by Grosberg and Strathmann (2007) as follows,

a   The compartmentalisation of replicating molecules, yielding the first cells.
b   The coalescence of replicating molecules to form chromosomes.
c   The use of DNA and proteins as the fundamental elements of the genetic code and replication.
d   The consolidation of symbiotic cells to generate the first eukaryotic cells containing chloroplasts and mitochondria.
e   Sexual reproduction involving the production (by meiosis) and fusion of haploid gametes.
f   The evolution of multicellular organisms from unicellular ancestors.
g   The establishment of social groups composed of discrete multicellular individuals.

Given the slow, progressive changes in this series of biological transitions, it seems noteworthy to underline the estimated time lapse in which they occurred, as summarised in Grosberg and Strathmann (2007). According to these authors, the first evidence of this transition comes from fossils of prokaryotic filamentous and mat-forming Cyanobacteria-like organisms, dating back 3.0–3.5 billion years. Cell differentiation would exceed 2 billion years ago, with multicellular eukaryotes existing perhaps 1 billion years ago, with a significant burst of metazoan diversification 600–700 Mya ago, during dramatic increases in atmospheric and oceanic oxygen. Thus, survival mechanisms of comparatively *simple* forms developed subjected to major ecosystem adaptations through a time-lapse of billion years. According to the authors mentioned above, in developmental terms, the transition from unicellular to multicellular organisation of bacteria and other organisms would be an inducible response to environmental stimuli.

Microbiome descendants of some of these survival experts managed to colonise newer, more complex life forms and became necessary for their emergence and survival. Today, mammalian guts represent one of Earth's significant reservoirs of microbial biodiversity (Thompson et al. 2017). Regarding the transition to multicellularity, the authors consider that only three lineages produced complex organisms, plants, animals, and fungi, in 3.5 billion years, i.e., not that many options.

> … there is now persuasive evidence that all extant eukaryotes are derived from an association with intracellular bacteria within the Rickettsiales that evolved into

mitochondria (Williams et al. 2007), with the implication that this propensity to form persistent associations has very ancient evolutionary roots.

(Douglas 2014)

Homo sapiens emerged in Natural History not from a *tabula rasa* condition but after an erratic, environmentally conditioned process of trials and errors of biological entities competing and sharing either as micro- or macro-species with variable ranges of survival capacities. Macro-species were hijacked by ubiquitous, comparatively simple living forms – survivors of aeons of critically changing environmental conditions – that colonised all significant ecological landscapes and involved organisms. These are, in our days, our microbiome partners and of several other living species, as some clades of animals are dependent on gut communities for ecological and evolutionary success. They represent the major survivors of our living planet and, so far, provide conditions for extending host evolutionary potential. As mentioned by Henry et al. (2021), under this hologenome concept, evolution operates on this single unit because eukaryotic hosts are never isolated from microbes in the natural world. According to this author,

> … work from community ecology, quantitative genetics, and evolutionary biology suggest that the microbiome frequently shapes host phenotypic distributions across taxa and environments and may play a critical role in host evolution.

It must be considered that the concept of holobiome has entered a debate on its usefulness. Hurst (2017) sustains that with direct heredity, the fitness of the symbiont is strongly correlated to that of the "host" individual. According to this author, in humans, twin studies provide evidence of a greater resemblance of gut microbiome constitution in identical twins compared to fraternal twins. Thus, the fate of the symbiont rests firmly on the fate of the individual who can carry and transmit it. The author adds,

> … it appears that the holobiont as a level of selection is, pragmatically, not central for understanding the diversity or evolution of holobionts. To understand the individual and its microbial community, we need to understand community assembly and the ecological dynamics of microbial communities, the role of evolution within the host in driving changes in the microbiome, the variation in diet/other conditions that may favour different strains.

Hurst (2017) further argues that,

> … microbial heredity – the direct passage of microbes from parent to offspring – is a key factor determining the degree to which the holobiont can usefully be considered a level of selection. Where direct vertical transmission (VT) is common, microbes form part of extended genomes whose dynamics can be modelled with simple population genetics …

In studies involving faecal microbiomes obtained from 1,126 twin pairs from the Twins UK registry, Goodrich et al. (2016) reported heritability and genome-wide association. According to these authors, heritability estimates were broadly similar when tripling the number of subjects from a previous report and the confidence intervals were narrower with the expanded dataset, making the estimates more robust. As stated by the authors, microbiome heritability would be slightly lower than those of other complex trait measures in the same population. Furthermore, diet, metabolism, and immune defence are essential drivers of human microbiome coevolution. In a previous study, Goodrich et al. (2014) reported that results represent strong evidence that the host's genetic makeup partly influences specific gut microbiota members. Further, they stated that the abundances of taxa are more highly correlated within monozygotic than dizygotic twin pairs. Thus, host genetic interactions with specific microbiome taxa are likely widespread across human populations with microbial phenotype influenced by human genetic state.

According to Dominguez-Bello et al. (2019),

> Rupture of the chorioamniotic membrane allows exposure of the baby to the maternal vaginal and perineal faecal microbes. Infants are naturally born with their skin and mouth covered by maternal inocula and have swallowed these microbes, supported by the observation of both DNA and live bacteria in the meconium.

Whether the holobiont (i.e., host plus microbiome) acts as a single evolutionary unit has also been challenged by several authors since microbes are essentially acquired horizontally and affected by multiple factors; additionally, there are significantly different spatial and temporal microbiome communities, which determines difficulties in expressing heritability in terms of hologenome (Douglas et al. 2020). As stated by these authors, by taking the hologenome model to be correct, it has been implied that microbial genetic variation should be integrated into host phenotype heritability estimates.

The holobiont could be defined as an evolutionary unit combining the eukaryotic host and associated microbes, with evolution operating on this single unit. Under this proposed construction, as stated by Henry et al. (2021), if the microbiome expands the host genetic repertoire and influences trait heritability, the microbiome may substantially impact host phenotypic evolution. However, these authors consider that the intertwined fate of a single host-microbiome evolutionary unit is the focus of much criticism on the hologenome concept.

One essential criticism is that the host and microbiome rarely operate as one selective unit because transmission of the microbiome between host generations is rarely strictly vertical. Douglas and Werren (2016) raised further arguments regarding the proposal of an evolutionary unit, questioning that host-microbe symbiosis does not imply a holobiont. According to these authors,

> … the hologenome is based on overly restrictive assumptions which render it an approach of little research utility. A host plus its microbiome is more effectively viewed as an ecological community of organisms that encompasses a broad range of interactions (parasitic to mutualistic), patterns of transmission (horizontal to vertical), and levels of fidelity among partners. The hologenome requires high partner fidelity if it is to evolve as a unit. However, even when this is achieved by

particular host-microbe pairs, it is unlikely to hold for the entire host microbiome, and therefore the community is unlikely to evolve as a hologenome… In contrast to the hologenome perspective, we argue here that the fields of ecology, genetics, and evolution are a perfectly adequate, and indeed more effective, conceptual framework for investigating the ecology and evolution of host-microbiome systems. In particular, a perspective that considers the host-microbiome as an ecological community.

Though these authors also stated that,

> … the question of whether host-microbe systems could evolve into units of higher biological organization is reasonable. In fact, two clear cases are the microbial ancestors of mitochondria and chloroplasts that evolved into organelles. However, it must be noted that these events are rare, and the rare examples involve individual host-symbionts becoming integrated into a "hologenome," not entire microbiomes.

Furthermore, Groussin et al. (2017) considered that,

> … as was recently pointed out, evidence for a tight association between a small number of taxa and their hosts does not necessarily generalize to the entire microbiome, and we find that only a minority of lineages fully co-speciate with their hosts. Moreover, evidence for co-speciation does not necessarily imply co-evolution, which needs to be established through more in-depth and mechanistic studies.

Additionally, on the debatable side, van Vliet and Doebeli (2019) considered the holobiont as a controversial concept based on the rationale that for selection to act at the holobiont level, there should be an association between the genotype of a host and the composition of its microbiome, that could result from vertical transmission (from the host to their offspring) –a condition met in mammals according to several authors as mentioned earlier and in Chapter 7. According to van Vliet and Doebeli (2019), most hosts constantly remove microbes from their environment, weakening the heritable association between a host and its microbiome. This has suggested that selection at the holobiont level is unlikely to play a significant role in nature. Their mathematical model would show that altruistic traits between microbes and host can evolve only under stringent conditions: hosts must pass on their microbiome to their offspring, and their generation time must be short, making it unlikely *in long-lived* mammals.

As discussed in Hurst (2017), despite discrepancies regarding the use of terms such as holobiont and hologenome, there is also a fundamental agreement in the need to understand the evolutionary ecology and dynamics of host–microbe interactions. Mayer et al. (2014), and Berg et al. (2020), considered that there has been a fundamental paradigm shift in our understanding of microorganisms, and it is now accepted that all eukaryotes are meta-organisms and must be considered together with their microbiota as an inseparable functional unit. In this context, it is important to consider host-microbe coevolution to facilitate a holistic understanding of the microbiota, with the microbiome concept defined as an ecological context, as a community of commensal, symbiotic, and pathogenic microorganisms within a body space or other environment (cf. Lederberg, 2001).

> Gut microbes may not provide the answers to all of the remaining questions we have regarding human ecology and evolution, but the potential magnitude of microbial contributions to human biology is too large to continue to be overlooked.
>
> *(Amato 2016)*

> ... coevolution of host-microbiota has likely shaped evolving phenotypes in all life forms on this predominantly microbial planet.
>
> *(Dominguez-Bello et al. 2019)*

Osmanovic et al. (2018) considered that evolution has been expanded to consider various non-Mendelian inheritance, including transmission of symbiotic microorganisms. The authors further expanded this concept, stating that the case of symbiotic organisms may be of particular interest because of its broad relevance to animals and plants and the potential of host-microbe interactions to support adaptations that were traditionally considered impossible for hosts and bacteria on their own. Additionally, the authors posed whether symbiosis between a host and microorganisms (holobiont) warrants a significant change in evolutionary thinking, which is currently under debate, as it is considered in sections of this text. However, for these authors, it is not clear whether the association between host and bacteria is tight enough to consider the holobiont as a unit of selection in evolution. Osmanovic et al. (2018) explored adaptation dynamics in a host-microbiome model in which the Darwinian selection of the host is coupled to a faster selection of its vertically transmitted bacteria. Their model relies on Darwinian selections operating on hosts and bacteria. According to the authors, interaction between these selections can give rise to previously unrealised modes of emergent adaptation promoted by bacterial influence on the survival probability of the host. This includes the selection of hosts based on a collective property of their bacterial community.

Thus, the intertwined fate of a single host microbiome evolutionary unit is the focus of debate regarding the hologenome. Though debate persists regarding the nature of host-microbiome relationships and their relationship with hereditary domains, the following excerpts provide a sample of conceptual standings,

> There is increasing evidence that the health benefits of symbiosis are commonly a consequence of microbial modulation of the signalling networks by which the growth and physiological function of eukaryote hosts are coordinated.
>
> *(Douglas 2014)*

> From a quantitative genetics perspective, heritability of the microbiome can be described as the proportion of microbial variation attributable to host genetic variance, just like any other complex trait. Thus, heritable microbes are those where the relative abundance or community structure is associated with particular host genotypes.
>
> *(Henry et al. 2021)*

*

In placing these considerations under an evolutive perspective it may be proper to include the following comments.

Let us consider that based on geological concepts, we are traversing the Anthropocene Era, according to some authors, what would be an additional Geological Era following the Cenozoic, Mesozoic, and Paleozoic of the Phanerozoic Eon, after the Proterozoic, Archean, and Hadean Aeons, as described in Figure 2.1. This covers approximately 4,000 million years, with the earliest forms of life appearing approximately at 3,500 million or earlier. During this appalling time scale, successive minor and major species disappeared – estimated at an extinction rate of 70–96% of species during a process of probably five mass extinction events profoundly affecting Earth biota, besides extinctions that would occur periodically at a "background rate" – and new ones emerged. As quoted in Ritchie (2022), there would be a natural background rate of the timing and frequency of extinctions: 10% of species are lost every million years, 30% every ten million years, and 65% every 100 million years. Archean, prokaryotes, and eukaryotes survived these significant and minor extinction events and continued with their "blind" opportunistic multiplication and colonisation drives that affected host evolutionary potential. Moeller and Sanders (2020) commented that mammalian hosts derive significant benefits from the gut microbiota. That is, gut microbial genomes expanded its metabolic potential relative to their hosts' genomes, some of which allow hosts to access foodstuffs that would be inaccessible otherwise.

In 2016, Alberdi et al. advanced the concept that although the host genome's differential gene expression and post-transcriptional processes received focus as the main molecular mechanisms, growing evidence points to the gut microbiota as a critical driver in defining hosts' phenotypes. As later considered by Henry et al. (2021), there would be two general ways the microbiome could affect host evolutionary potential. That is, either by shifting the mean host phenotype or changing the variance in host phenotypes in the population. Thus, the survival and adaptive potential of comparatively simpler microorganisms allowed them to survive and recreate association processes with complex macro-organisms that, in their complexity, turned out to be more fragile or less adaptive to major ecological events.

The survival and ubiquitous capacity of Archean, prokaryotes, and eukaryotes define the persistent survival profile of Natural History and provide a starkly different evolution of the traditional view of human biological construction and behavioural processes. As stated by Henry et al. (2021),

> Despite their clear importance, the complex effects of microbial inheritance and genetics on host phenotypes remain underappreciated, limiting our ability to understand host-microbiome evolution.

The previous quotation implies a conceptual limitation, as mentioned by Henry et al. (2021), that ought to be further explored to involve our current, prevailing concepts regarding our human nature, relative freedom, and dominant standing among the multitude of living forms in Natural History. Living creatures and ecosystems form a dynamic web of interactive nature, where the construction and expression of individual profiles emerge conditioned by such a web. *Homo sapiens* manage to add to it a species variant – a sort of indentation in such a web – primarily based on its creativeness that provides a virtual sense or flavour of individual freedom that goes beyond our real biological construction. Freedom is further conditioned by human social constructions that generate rampant inequities among social groups (Colombo 2023).

Despite this proposal of a holistic view of the multiple conditioning interactions in Nature, the progressive evolution of the successive brain and mental complexities in the *Homo* species suggests that its evolution operates on stratified levels with inter-active connections. To what extent is our species' creativity devoid of, or immune to, influences and conditioning from lower processing stages immersed in the larger web of interactions among micro- and macro-organisms? A series of reports based on behavioural and clinical contexts suggest that we would rather respond to a complex web of mutual dependence, as is summarised in a later chapter.

References

Alberdi, Antton, *et al.* "Do Vertebrate Gut Metagenomes Confer Rapid Ecological Adaptation?" *Trends in Ecology & Evolution*, vol. 31, no. 9, 2016, pp. 689–699, doi:10.1016/j.tree.2016.06.008.

Amato, Katherine R. "Incorporating the Gut Microbiota into Models of Human and Non-Human Primate Ecology and Evolution." *American Journal of Physical Anthropology*, vol. 159, no. S61, 2016, pp. 196–215, doi:10.1002/ajpa.22908.

Berg, Gabriele, *et al.* "Microbiome Definition Re-Visited: Old Concepts and New Challenges." *Microbiome*, vol. 8, no. 1, 2020, doi:10.1186/s40168-020-00875-0.

Bocci, Velio. "The Neglected Organ: Bacterial Flora has a Crucial Immunostimulatory Role." *Perspectives in Biology and Medicine*, vol. 35, no. 2, 1992, pp. 251–260, doi:10.1353/pbm.1992.0004.

Burokas, Aurelijus, *et al.* "Microbiota regulation of the Mammalian Gut–Brain Axis." *Advances in Applied Microbiology*, vol. 91, 2015, pp. 1–62, doi:10.1016/bs.aambs.2015.02.001.

Colombo, Jorge A. *Evolution and the Human-Animal Drive to Conflict: A Psychobiological Perspective*. Abingdon: Routledge, 2023, doi:10.4324/9781003387695.

Dawkins, R. *The Selfish Gene*. Oxford: Oxford University Press, 1989.

Dominguez-Bello, Maria Gloria, *et al.* "Role of the Microbiome in Human Development." *Gut*, vol. 68, no. 6, 2019, pp. 1108–1114, gut.bmj.com/content/68/6/1108, doi:10.1136/gutjnl-2018-317503.

Douglas, Angela E. "Symbiosis as a General Principle in Eukaryotic Evolution." *Cold Spring Harbor Perspectives in Biology*, vol. 6, no. 2, 2014, pp. a016113–a016113, cshperspectives.cshlp.org/content/6/2/a016113.full, doi:10.1101/cshperspect.a016113.

Douglas, Angela E., and John H. Werren. "Holes in the Hologenome: Why Host-Microbe Symbioses Are Not Holobionts." *mBio*, vol. 7, no. 2, 2016, doi:10.1128/mbio.02099-15.

Douglas, Gavin M., *et al.* "Re-Evaluating the Relationship between Missing Heritability and the Microbiome." *Microbiome*, vol. 8, no. 1, 2020, doi:10.1186/s40168-020-00839-4.

Foster, Kevin R., *et al.* "The Evolution of the Host Microbiome as an Ecosystem on a Leash." *Nature*, vol. 548, no. 7665, 2017, pp. 43–51, doi:10.1038/nature23292.

Goodrich, Julia K., *et al.* "Human Genetics Shape the Gut Microbiome." *Cell*, vol. 159, no. 4, 2014, pp. 789–799, doi:10.1016/j.cell.2014.09.053.

Goodrich, Julia K., *et al.* "Genetic Determinants of the Gut Microbiome in UK Twins." *Cell Host & Microbe*, vol. 19, no. 5, 2016, pp. 731–743, doi:10.1016/j.chom.2016.04.017.

Grosberg, Richard K., and Richard R. Strathmann. "The Evolution of Multicellularity: A Minor Major Transition?" *Annual Review of Ecology, Evolution, and Systematics*, vol. 38, no. 1, 2007, pp. 621–654, doi:10.1146/annurev.ecolsys.36.102403.114735.

Groussin, Mathieu, *et al.* "Unraveling the Processes Shaping Mammalian Gut Microbiomes Over Evolutionary Time." *Nature Communications*, vol. 8, no. 1, 2017, doi:10.1038/ncomms14319.

Henry, Lucas P., *et al.* "The Microbiome Extends Host Evolutionary Potential." *Nature Communications*, vol. 12, no. 1, 2021, p. 5141, www.nature.com/articles/s41467-021-25315-x, doi:10.1038/s41467-021-25315-x.

Huitzil, Saúl, *et al.* "Modeling the Role of the Microbiome in Evolution." *Frontiers in Physiology*, vol. 9, 2018, doi:10.3389/fphys.2018.01836.

Hurst, Gregory D. D. "Extended Genomes: Symbiosis and Evolution." *Interface Focus*, vol. 7, no. 5, 2017, www.ncbi.nlm.nih.gov/pmc/articles/PMC5566813/, doi:10.1098/rsfs.2017.0001.

Ley, Ruth E., *et al.* "Ecological and Evolutionary Forces Shaping Microbial Diversity in the Human Intestine." *Cell*, vol. 124, no. 4, 2006, pp. 837–848, doi:10.1016/j.cell.2006.02.017.

Luckey, T. D. "Introduction to Intestinal Microecology." *The American Journal of Clinical Nutrition*, vol. 25, no. 12, 1972, pp. 1292–1294, doi:10.1093/ajcn/25.12.1292.

Mayer, Emeran A., *et al.* "Gut Microbes and the Brain: Paradigm Shift in Neuroscience." *Journal of Neuroscience*, vol. 34, no. 46, 2014, pp. 15490–15496, www.jneurosci.org/content/34/46/15490.short, doi:10.1523/jneurosci.3299-14.2014.

McFall-Ngai, Margaret, *et al.* "Animals in a Bacterial World, a New Imperative for the Life Sciences." *Proceedings of the National Academy of Sciences*, vol. 110, no. 9, 2013, pp. 3229–3236, doi:10.1073/pnas.1218525110.

Moeller, Andrew H., and Jon G. Sanders. "Roles of the Gut Microbiota in the Adaptive Evolution of Mammalian Species." *Philosophical Transactions of the Royal Society B: Biological Sciences*, vol. 375, no. 1808, 2020, p. 20190597, doi:10.1098/rstb.2019.0597.

Moran, Nancy A., *et al.* "Evolutionary and Ecological Consequences of Gut Microbial Communities." *Annual Review of Ecology, Evolution, and Systematics*, vol. 50, no. 1, 2019, pp. 451–475, doi:10.1146/annurev-ecolsys-110617-062453.

Münger, Emmanuelle *et al.* "Reciprocal Interactions between Gut Microbiota and Host Social Behavior." *Frontiers in Integrative Neuroscience*, vol. 12, 2018, doi:10.3389/fnint.2018.00021.

O'Hagan, Caroline, *et al.* "Long-Term Multi-Species *Lactobacillus* and *Bifidobacterium* Dietary Supplement Enhances Memory and Changes Regional Brain Metabolites in Middle-Aged Rats." *Neurobiology of Learning and Memory*, vol. 144, 2017, pp. 36–47. doi:10.1016/j.nlm.2017.05.015.

O'Hara, Ann M., and Fergus Shanahan. "The Gut Flora as a Forgotten Organ." *EMBO Reports*, vol. 7, no. 7, 2006, pp. 688–693. doi:10.1038/sj.embor.7400731.

Osmanovic, Dino, *et al.* "Darwinian Selection of Host and Bacteria Supports Emergence of Lamarckian-like Adaptation of the System as a Whole." *Biology Direct*, vol. 13, no. 1, 2018, doi:10.1186/s13062-018-0224-7.

Risely, Alice. "Applying the Core Microbiome to Understand Host-Microbe Systems." *Journal of Animal Ecology*, vol. 89, no. 7, 2020, pp. 1549–1558, doi:10.1111/1365-2656.13229.

Ritchie, Hannah. "There Have Been Five Mass Extinctions in Earth's History." *Our World in Data*, 2022, ourworldindata.org/mass-extinctions.

Rosenberg, Eugene, and Ilana Zilber-Rosenberg. "The Hologenome Concept of Evolution after 10 Years." *Microbiome*, vol. 6, no. 1, 2018, doi:10.1186/s40168-018-0457-9.

Sarkar, Amar, *et al.* "The Role of the Microbiome in the Neurobiology of Social Behaviour." *Biological Reviews*, vol. 95, no. 5, 2020, pp. 1131–1166, doi:10.1111/brv.12603.

Shapira, Michael. "Gut Microbiotas and Host Evolution: Scaling up Symbiosis." *Trends in Ecology & Evolution*, vol. 31, no. 7, 2016, pp. 539–549, doi:10.1016/j.tree.2016.03.006.

Sharp, Connor, and Kevin R. Foster. "Host Control and the Evolution of Cooperation in Host Microbiomes." *Nature Communications*, vol. 13, no. 1, 2022, doi:10.1038/s41467-022-30971-8.

She, Xinwei, *et al.* "Shotgun Sequence Assembly and Recent Segmental Duplications within the Human Genome." *Nature*, vol. 431, no. 7011, 2004, pp. 927–930, doi:10.1038/nature03062.

Slyepchenko, Anastasiya, *et al.* "Gut Emotions - Mechanisms of Action of Probiotics as Novel Therapeutic Targets for Depression and Anxiety Disorders." *CNS & Neurological Disorders – Drug Targets*, vol. 13, no. 10, 2015, pp. 1770–1786, doi:10.2174/1871527313666141130205242.

Thompson, Luke R., *et al.* "A Communal Catalogue Reveals Earth's Multiscale Microbial Diversity." *Nature*, vol. 551, no. 7681, 2017, pp. 457–463, doi:10.1038/nature24621.

van Vliet, Simon, and Michael Doebeli. "The Role of Multilevel Selection in Host Microbiome Evolution." *Proceedings of the National Academy of Sciences*, vol. 116, no. 41, 2019, pp. 20591–20597, doi:10.1073/pnas.1909790116.

Wong, Adam C-N., *et al.* "Behavioral Microbiomics: A Multi-Dimensional Approach to Microbial Influence on Behavior." *Frontiers in Microbiology*, vol. 6, 2015, doi:10.3389/fmicb.2015.01359.

Zhang, Jie, *et al.* "Probiotic Has Prophylactic Effect on Spatial Memory Deficits by Modulating Gut Microbiota Characterized by the Inhibitory Growth of *Escherichia Coli.*" *Frontiers in Integrative Neuroscience*, vol. 17, 2023, doi:10.3389/fnint.2023.1090294.

Zilber-Rosenberg, Ilana, and Eugene Rosenberg. "Role of Microorganisms in the Evolution of Animals and Plants: The Hologenome Theory of Evolution." *FEMS Microbiology Reviews*, vol. 32, no. 5, 2008, pp. 723–735, doi:10.1111/j.1574-6976.2008.00123.x.

5

MICROBIOTA AND SPECIES EVOLUTION

Host-Microbiome Evolution and Human Microbiome

Our gut microbiota can be pictured as a microbial organ placed within a host organ: It is composed of different cell lineages with a capacity to communicate with one another and the host; it consumes, stores, and redistributes energy; it mediates physiologically important chemical transformations; and it can maintain and repair itself through self-replication.

(Bäckhed et al. 2005)

The microorganisms that live inside and on humans (known as the microbiota) are estimated to outnumber human somatic and germ cells by a factor of ten. Together, the genomes of these microbial symbionts (collectively defined as the microbiome) provide traits that humans did not need to evolve on their own1. If humans are thought of as a composite of microbial and human cells, the human genetic landscape as an aggregate of the genes in the human genome and the microbiome, and human metabolic features as a blend of human and microbial traits, then the picture that emerges is one of a human "supraorganism".

(The Human Microbiome Project, *Turnbaugh et al. 2007*)

The intestinal microbiota consists of a community of bacteria that colonize the gastrointestinal tract after birth and persist throughout adult life, and "transient" bacteria, such as probiotic bacteria, which are temporarily acquired during ingestion of certain foods.

(Bercik et al. 2012)

The microbiota, comprising communities of bacteria, viruses, fungi, and other microorganisms that live mutualistically in and on animals, is increasingly recognized as an essential component of normal physiology, with important roles in

DOI: 10.4324/9781032698380-5

health and disease. As the first life forms on the planet, microorganisms are integrated fundamentally across biological scales. They regulate the biogeochemical cycling of elements essential for life; form the likely endosymbiotic origins of genomic elements, eukaryotic organelles, and multicellular organisms; and maintain homeostatic interactions within and across plant, animal, and atmospheric ecosystems.

(Vuong et al. 2017)

To place animal emergence on comparative time grounds with bacterial emergence, as stated by McFall-Ngai et al. (2013), animals diverged from their protistan ancestors 700–800 Mya ago, some 3 billion years after bacterial life originated and as much as 1 billion years after the first appearance of eukaryotic cells. Thus, an ancient repertoire of eukaryote–bacterial interactions was established.

The conundrum posed by the events leading to our species' emergence and its behavioural complexity and evolution would have probably occurred in multiple genetic steps that must have left detectable footprints in our genomes, as posed by Gagneux and Varki (2001). Varki and Altheide (2005) stated that the difference between human and chimpanzee genomes would vary by approximately 4%. According to Patterson et al. (2006), the genetic divergence time between humans and chimpanzees would significantly vary across the genome, conveying important information about the timing and process of speciation since both lineages could have exchanged genes before separating permanently. In this domain, what is not known is the extent to which the ancestral population that gave rise to *anatomically modern humans* was genetically isolated and whether archaic *hominins* made a genetic contribution to the modern human gene pool, as proposed by Hammer et al. (2011). In other words, as stated by the authors, would our genes descend from divergent ancestors occupying different ecological niches over a more comprehensive geographical range across and outside the African Pleistocene landscape? These authors further state that polymorphisms present in extant populations are introgressed (i.e., transfer of genetic material from one species into the gene pool of another) via relatively recent interbreeding with hominin forms that diverged from the ancestors of modern humans in the Lower-Middle Pleistocene.

Thus, our species' origin would be a product involving genetic components expressed in different previous primate species. Host-gut microbiome interactions adapted to different feeding ecosystems provide an additional component to this genetic admixture.

The host coevolved with its intestinal microbiota throughout evolution, involving reciprocal adaptation and benefits. The microbiota, or microbiome, comprises communities of bacteria, viruses, fungi, and other microorganisms that live mutualistically in and on animals and compose an ecologically dynamic domain. According to Hurst (2017), experimental evolution studies have demonstrated that the large population size and short generation time of microbes make them labile on evolutive grounds and dependent on diet shifts by the host; hence, the microbiome is ecologically and evolutively dynamic.

The universe of microbes colonising the human digestive tract includes an estimated number of 150 to 400 species, with their number and relative proportion varying according to several conditioning variables, such as individual, age, ethnicity, and feeding habits. According to Qin et al. (2010), 99.1% of the genes in the gut microbiome catalogue are of bacterial origin, the remainder being mostly archaeal, with only 0.1% of eukaryotic and viral origins. As Bruijning et al. (2020) discussed, the benefits and stability of a given host-microbiome association will ultimately be a function of the environmental context, with different expectations for hosts that inhabit stable versus more rapidly fluctuating environments. These authors further stated that imperfect transmission may be adaptive by allowing individuals to ensure phenotypic variability in their offspring in contexts where varying environments mean this strategy increases long-term fitness.

Based on previous reports, Ley et al. (2006) considered that the intestine is remarkable for its exclusivity, adding that the microbial community presumably has strict requirements for membership. Selection pressures trim the microbial diversity of the outside world so that the gut microbiota in adults would be dominated by members of only two divisions of bacteria – the *Bacteroidetes* and *Firmicutes*– and one member of Archaea, the *Methane brevibacter smithii*. According to these authors, natural selection forces selection pressure to drive microbial cells to become functionally specialised, while host selection on the community favours stable societies with a high degree of functional redundancy. Walter and Ley (2011) reported that each millilitre of the large intestine holds approximately 10^{11} microbial cells compared to 10^8 cells in the small intestine.

According to Van den Abbeele et al. (2011), the host selects the intestinal microbiota, particularly the mucosa-associated microbial community which depends on diet habits. While mucosal microorganisms would be crucial for immunological priming, luminal microorganisms would be essential for nutrient digestion. In this regard, the authors remarked that the host has developed a mucosal defence barrier along the intestinal tract. The mucus is constantly shed off from the epithelium, and intercellular tight junctions form a virtually impermeable barrier for molecules and ions, allowing the host to absorb specific compounds or antigens through controlled mechanisms.

As mentioned by Davenport et al. (2017), besides ancestral cultural feeding habits, Western microbiomes differ in numbers and composition from the non-Western microbiomes profiled to date, most clearly between industrialised and non-industrialised countries. The numerical dimension is daunting, considering that the intestinal microbiota community harbours by far the greatest microbial biomass of any organ or surface of the human body.

According to Kurilshikov et al. (2017), human and mouse studies have thus far shown that a subset of gut bacteria is heritable and influenced by host genetics. Consequences of changes in diet habits on the gut microbiome are exemplified by Prescott et al. (2016), who comment on data showing that recent research from North America demonstrates reduced diversity of colonic microbiota among residents of lower socioeconomic status (SES). An opposite pattern can occur in developing nations where lower SES is associated with microbial diversity. The impact of this variable was also studied in a United Kingdom twins cohort exposed to a different set of deprivations (Bowyer et al. 2019). The authors concluded that they significantly differ by measures of microbiome compositional dissimilarities. Additionally, the more significant the difference in the twin pair Index of Multiple Deprivation, the greater the dissimilarity of their microbiota will be.

Though it has been characterised as what could be considered core microbiota (Gupta et al. 2017), these authors described concurrent differences in the microbiome with a shift in individual samples from living styles of subsistence like foraging, rural farming, and industrialised urban western life.

> We share our body space with around 100 trillion *microorganisms*, collectively known as microbiota (Turnbaugh *et al.*, 2007; Consortium, 2012). The growing perception that *our* genetic landscape is a summation of the genes embedded *in* our own genome as well as in genomes of our microbiota (the microbiome), and that our metabolic features present an assemblage of human and microbial traits …
>
> *(Gupta et al. 2017)*

*

> As one of the earliest researchers focusing on the far-reaching consequences of alterations to intestinal microbiota, Dubos was also mindful of the place of microbes in the ecosystems of life (1963): "Animals and man have evolved in intimate association with a complex microbial flora. It is to be expected therefore, that many characteristics of their anatomical and physiological development reflect this evolutionary past and are the manifestations of tissue responses to the microflora … life can be viewed through the lens of the holobiont, that is, the multicellular eukaryote and the inseparable colonies of persistent symbionts which together form a critically important unit of anatomy, physiology, immunology, growth, and evolution. At the very least, the host plus its microbiome (microbiota and their collective genomes) can be regarded as an ecological community."
>
> *(Prescott et al. 2016)*

> The evolution of speech likely aided in the ability of protohumans to coordinate the behaviour of the group. It has recently become clear that bacteria made this discovery at least a billion years earlier, and in lieu of speech, they evolved a rich lexicon of diffusible chemical signals. In the past decade we have learned that bacteria use these signals to communicate both within and between species, and they also use a wide variety of chemical signaling molecules, signal detection apparatuses, and signal transduction mechanisms.
>
> *(Winans & Bassler 2002)*

Though species construct group behavioural patterns, codes of interaction, and hierarchical structures – a sort of protoculture, or evolutive prologue to the complex build-up of the human cultural construction – this process occurs mounted on biological interaction among several commensals. These are defined according to different levels of their interaction with the ecosystem. As characterised by Moran et al. (2019), hosts may evolve dependence on the presence of gut bacteria such that removal of the normal gut microbiota represents an abnormal environment that leads to abnormal development or behaviour and usually to lowered fitness. Alternatively, they may confer new capabilities, resulting in ecological expansion and evolutionary diversification.

From an evolutive perspective, microscopic commensals are *surrogates* of the primal biological entities of all species, as outlined in previous chapters. As stated by Vuong et al. (2017), since microorganisms represent the first lifeforms, they are essentially integrated across biological scales, within and across plants, animals, and ecosystems.

As it was mentioned earlier, animals diverged from their protistan ancestors 700–800 Mya, some 3 billion years after bacterial life originated and as much as 1 billion years after the first appearance of eukaryotic cells, becoming integrated units of micro and macro systems adopting diverse forms of sharing and commensalism of diverse complexity.

> … the current-day relationships of protists with bacteria, from predation to obligate and beneficial symbiosis, were likely already operating when animals first appeared.
>
> *(McFall-Ngal et al. 2013)*

This comment underscores the ancient interdependence among complex multicellular organisms and associated microorganisms. These interactions developed as evolution generated higher complex organisms, which McFall-Ngai et al. (2013) stressed with the development of progressively complex tubular digestive systems and their associated microbiota.

Apparently, with large variability among insects (Sanders et al. 2017), the animal kingdom shares a community of bacteria located in outer/inner (gut) structures. The microbial-sharing universe includes cold-water corals (Galand et al. 2019). A comparative analysis of microbiome in non-human animals has been conducted by Hanning and Diaz-Sanchez (2015), with analysis among wild and captive animals studied by de Jonge et al. (2022).

According to the *Human Microbiome Project Consortium* (NIH HMP) (2012)[1], the human body contains trillions of microorganisms outnumbering human cells by ten to one, though this ratio has been challenged by Sender et al. (2016), who updated the ratio of bacteria to human cells in the body from 10:1 or 100:1, to closer to 1:1. Because of their small size, however, microorganisms make up only about 1 to 3% of the body's mass. The NIH HMP (2012) researchers also reported that this plethora of microbes contributes more genes responsible for human survival than humans. Where the human genome carries some 22,000 protein-coding genes, researchers estimate that the human microbiome contributes some 8 million unique protein-coding genes or 360 times more bacterial genes than human genes. The complex ecosystem of the microbiome – composed of thousands of species of bacteria, viruses, fungi, and protozoa – and its genomic[2] contribution and host interaction are critical for human survival as they affect human metabolic and immune function, as well as behaviour, as mostly based on experimental evidence. In this regard, Moeller and Sanders (2020), based on Xiao et al. (2015) data obtained from mice, quote that whereas the typical mammalian genome contains 10,000 genes, typical mammalian gut microbiota can contain several million evolutionarily distinct genetic families.

Xiao et al. (2015) stated that the mouse gut microbiome is functionally like its human counterpart, with 95.2% of its Kyoto Encyclopaedia of Genes and Genomes (KEGG) orthologous groups in common. However, only 4.0% of the mouse gut microbial genes were shared (95% identity, 90% coverage) with those of the human gut microbiome. As commented by Org et al. (2015) and Kurilshikov et al. (2017), human and mouse studies have thus far shown that a subset of gut microbiota bacteria would be heritable and influenced by host genetics and environment. Moeller and Sanders (2020) considered that the dietary practices of current humans may have consequences for the gut microbiota of future generations. These authors quote studies consistent with this possibility, according to which humans living in industrialised societies, on average, would harbour lower levels of gut microbial diversity than any other wild-living primates.

These comments frame future considerations regarding the *cultural onion layers* mounted on top of bacterial commensality and ancient behavioural drives, as discussed in Colombo (2019, 2022). Therefore, one fundamental question would be the extent to which animal microbiome would affect evolution and behaviour before addressing its impact on our species. Or, as posed by Blaser and Falkow (2009), what would be the consequences of human microbiota disappearance?

Regarding its possible impact on brain development of *Homo sapiens*, several hypotheses and theories have been proposed. Among them, social complexity as an ecological factor (Dunbar 2009; Dunbar and Shultz, 2007), shift in feeding behaviour with the subsequent reduction of the digestive system to compensate for energy demand due to brain growth (Aiello and Wheeler 1995); genetic modifications (see before); neurotransmitter involvement, such as dopamine (Previc 1999, 2006), and culture (see ahead). These proposals do not seem to contradict but rather complement each other, acting at similar or different (sequential) times.

As it will be further dealt with later, the gut microbiome represents a critical factor in the multifactorial equation involved in *sapiens* or *Homo* development. Its role in human physiologic integration would suggest that the human organism represents a multi-genomic entity. Hence, besides our DNA, a universe of bacteria coexists in human organisms. According to Ley et al. (2008a), diet and phylogeny regulate the gut microbiome. Its diversity increases from carnivores to omnivores to herbivores, and humans would be typical of omnivore primates. On a metabolic dimension, the gut microbiome would codify functions that our genome would not have developed entirely on its own. This would include the capacity to extract nutrients and energy from our diet according to Ley et al. (2008a). Gut microbiome would represent a potentially significant factor among those involved in species evolution.

According to Buchanan and Bohórquez (2018), the relationship of the gut with developmental issues precedes postnatal life. Circulation of amniotic fluid components after being swallowed by the foetus is the first medium interacting with the prenatal gut. In this respect, it seems worth mentioning that the human amniotic fluid has been shown to foster the growth of rat neural brain cells in culture. This effect is disrupted by antibodies to nerve and epidermal growth factors, as explored by Colombo et al. (1993). The increased permeability of gut walls at early stages of foetal development reported by Robertson et al. (1982) and Weaver et al. (1984) could allow amniotic components to enter the humoral media and act as a trophic brain component.

The microbiome establishes a close symbiotic relationship with the host since birth, as it will be further analyzed, although it goes through successive changes due to shifts in feeding habits as we grow. According to Borre et al. (2014a), feeding sequence and conditions impact development and adult behaviour. Experimentally, in laboratory animals, it has been possible to confirm that the microbiome could modulate behaviour as it may relate to social interaction, exploratory activity, and anxiety (Pinto-Sánchez et al. 2017), as well as affect the expression of neurotransmitters (GABA) and *brain-derived neurotrophic factor* (BDNF), according to Heijtz et al. (2011); Bercik et al. (2011); Bravo et al. (2011); Borre et al. (2014a) and Buffington et al. (2016).

The studies mentioned address the impact of the gut microbiome on the structure, physiology, and development of the nervous system and behaviour. In this context – feeding, gut microbiome, evolution – it is interesting to speculate on the possible role of feeding habits in the evolution of the *Homo* species. The microbiome existence explains the rise and fall of several common diseases in developed countries since it is considered that animal species share a *bacterial world* that includes complex endo-symbiotic relationships. "Disappearing microbiota" has been considered a factor in the maladaptation and emergence of certain diseases. Modern humans and their ancestors have evolved since the most ancient times with a commensal microbiota; more yet, it has been calculated that animals have carried resident microorganisms since at least the emergence of sponges, as quoted in Blaser and Falkow (2009). This involved coevolution, coadaptation, and co-dependency with several gains. These include facilitating energy extraction from food, providing accessory growth factors, promoting post-natal terminal differentiation of mucosal structure and function, stimulating innate and adaptive immune systems, and providing "colonisation resistance" against pathogen invasion. Other involvements of the microbiome in host physiology and behaviour have been referred to before. Thus, as Blaser (2006) mentioned,

> The existence of an indigenous—or residential—microbiota is ancient. Such microbes are found in all animals that are at least as complex as annelids (earthworms); this provides evidence for the co-evolution of animals and bacteria for more than 800 million years (Schramm et al, 2003, Savage, 1977). The microbiota at specific sites in our bodies and those of our mammalian relatives have many similarities in composition and organization, which is consistent with their common ancestry and/or convergent evolution.

As mentioned earlier, the first bacteria evolved over 3 billion years ago and dominated the biosphere continually, shaping the environment where animals would eventually evolve more than 2 billion years later. Early marine organisms evolved in close association with bacteria found in stromatolites (layered macroscopic sedimentary formations). This implied recording their ancient existence, and whether predating on them or harbouring bacterial commensals, they influenced animal origins. As discussed in Alegado and King (2014),

> Although the exact timing of eukaryotic origins is unknown, macroscopic structures found in Gabon, West Africa have been interpreted as eukaryotic fossils that are at least 2.1 billion years old (El Albaniet al. 2010). Undisputed multicellular eukaryotes did not appear in the fossil record until ⊠1.2 billion years ago (Butterfield 2000). Animals apparently lagged even further behind ...

*

The evolution of oxygen-dependent organisms was probably delayed by the low oxygen concentration and diffusion in the organismal cell layers until cyanobacteria initiated a terrestrial oxygen factory. Critical levels for living creatures probably were attained 2.3 billion years ago (Kasting and Siefert 2002; Canfield et al. 2013). According to Heijtz (2011), animals diverged from their protistan ancestors 700–800 Mya, some 3 billion years after bacterial life originated and as much as 1 billion years after the first appearance of eukaryotic cells. Since atmospheric evolution on an inhabited planet is determined mainly by its microbial populations, these primal circumstances conditioned the evolution. They did so by sharing the biological systems of more complex organisms, thus progressively affecting the conditions for their evolution and competing successfully in the prey-predatory domain through behavioural conditioning.

Thus, animal evolution proceeded as a host under a shared umbrella with bacterial communities (microbiome). As stated by Zoetendal et al. (2006), there is a microbial world within us. Regarding the impact of evolution on microbiota in primate species domains, according to Ochman et al. (2010),

> Our analyses revealed a clear species-specific signature of microbial community structure. Moreover, the pattern of relationships among the five great ape species (*Homo sapiens, Pan troglodytes, P. paniscus, Gorilla gorilla, and G. beringei*) inferred from their fecal microbial communities was identical to that inferred from host mitochondrial DNA, indicating that host phylogeny shapes the gut microbiota over evolutionary timescales.

According to these authors, host phylogeny significantly diversifies distal gut microbial communities in great apes.

There is a disagreement regarding the assertion that the human body contains trillions of microorganisms – outnumbering human cells by ten to one – since the mentioned ratio of bacteria to human cells has been rebutted by studies yielding almost equal numbers, as stated by Sender et al. (2016) and commented on by Abbott (2016). Human Microbiome Project researchers also reported that the number of microbes contributes more genes responsible for human survival than humans contribute. Thus, this bacterial genomic contribution would be critical for human survival.

Hidden under our cultural envelope and collectively ignored by our species' pride, macro-biological life generally thrives under a shared host-gut microbiome umbrella.

Notes

1 https://www.nih.gov/news-events/news-releases/nih-human-microbiome-project-defines-normal-bacterial-makeup-body.
2 https://www.nih.gov/news-events/news-releases/nih-human-microbiome-project-defines-normal-bacterial-makeup-body.

References

Abbott, Alison. "Scientists Bust Myth That Our Bodies Have More Bacteria than Human Cells." *Nature*, 2016, doi:10.1038/nature.2016.19136.

Aiello, Leslie C., and Peter Wheeler. "The Expensive-Tissue Hypothesis: The Brain and the Digestive System in Human and Primate Evolution." *Current Anthropology*, vol. 36, no. 2, 1995, pp. 199–221, doi:10.1086/204350.

Alegado, R. A., and N. King. "Bacterial Influences on Animal Origins." *Cold Spring Harbor Perspectives in Biology*, vol. 6, no. 11, 2014, doi:10.1101/cshperspect.a016162.

Bäckhed, F.*et al.* "Host-Bacterial Mutualism in the Human Intestine." *Science*, vol. 307, no. 5717, 2005, pp. 1915–1920, doi:10.1126/science.1104816.

Bercik, Premysl, *et al.* "The Intestinal Microbiota Affect Central Levels of Brain-Derived Neurotropic Factor and Behavior in Mice." *Gastroenterology*, vol. 141, no. 2, 2011, doi:10.1053/j.gastro.2011.04.052.

Bercik, P., *et al.* "Microbes and the Gut-Brain Axis." *Neurogastroenterology & Motility*, vol. 24, no. 5, 2012, pp. 405–413, doi:10.1111/j.1365-2982.2012.01906.x.

Blaser, Martin J. "Who are we? Indigenous Microbes and the Ecology of Human Diseases" *EMBO Reports*, vol. 7, no. 10, 2006, pp. 956–960, doi:10.1038/sj.embor.7400812.

Blaser, Martin J., and Stanley Falkow. "What are the Consequences of the Disappearing Human Microbiota?" *Nature Reviews Microbiology*, vol. 7, no. 12, 2009, pp. 887–894, doi:10.1038/nrmicro2245.

Borre, Yuliya E., *et al.* "Microbiota and Neurodevelopmental Windows: Implications for Brain Disorders." *Trends in Molecular Medicine*, vol. 20, no. 9, 2014, pp. 509–518, doi:10.1016/j.molmed.2014.05.002.

Bowyer, Ruth, *et al.* "Socioeconomic Status and the Gut Microbiome: A TwinsUK Cohort Study." *Microorganisms*, vol. 7, no. 1, 2019, doi:10.3390/microorganisms7010017.

Bravo, Javier A., *et al.* "Ingestion of *Lactobacillus* Strain Regulates Emotional Behavior and Central GABA Receptor Expression in a Mouse via the Vagus Nerve." *Proceedings of the National Academy of Sciences*, vol. 108, no. 38, 2011, pp. 16050–16055, doi:10.1073/pnas.1102999108.

Bruijning, Marjolein, *et al.*When the Microbiome Defines the Host Phenotype: Selection on Vertical Transmission in Varying Environments, 2020, bioRxiv doi: doi:10.1101/2020.09.02.280040.

Buchanan, Kelly L., and Diego V. Bohórquez. "You Are What You (First) Eat." *Frontiers in Human Neuroscience*, vol. 12, 2018, doi:10.3389/fnhum.2018.00323.

Buffington, Shelly A., *et al.* "Microbial Reconstitution Reverses Maternal Diet-Induced Social and Synaptic Deficits in Offspring." *Cell*, vol. 165, no. 7, 2016, pp. 1762–1775, doi:10.1016/j.cell.2016.06.001.

Canfield, Donald E., *et al.* "Oxygen Dynamics in the Aftermath of the Great Oxidation of Earth's Atmosphere." *Proceedings of the National Academy of Sciences*, vol. 110, no. 42, 2013, pp. 16736–16741, doi:10.1073/pnas.1315570110.

Colombo, Jorge A., *et al.* "Trophic Influences of Human and Rat Amniotic Fluid on Neural Tube-Derived Rat Fetal Cells." *International Journal of Developmental Neuroscience*, vol. 11, no. 3, 1993, pp. 347–355, doi:10.1016/0736-5748(93)90006-y.

Colombo, Jorge A. *Our Animal Condition and Social Construction*. New York: Nova Science Publishers Inc., 2019.eBook ISBN: 978-971-53615-53583.

Colombo, Jorge A. *Dominance Behavior: An Evolutive and Comparative Perspective*. Cham: Springer International Publishing, 2022.

Davenport, Emily R., *et al.* "The Human Microbiome in Evolution." *BMC Biology*, vol. 15, no. 1, 2017, doi:10.1186/s12915-017-0454-7.

de Jonge, Nadieh, *et al.* "The Gut Microbiome of 54 Mammalian Species." *Frontiers in Microbiology*, vol. 13, 2022, p. 886252, pubmed.ncbi.nlm.nih.gov/35783446/, doi:10.3389/fmicb.2022.886252.

Dunbar, Robin I. M., and Susanne Shultz. "Evolution in the Social Brain." *Science*, vol. 317, no. 5843, 2007, pp. 1344–1347, doi:10.1126/science.1145463.

Dunbar, Robin I. M. "The Social Brain Hypothesis and Its Implications for Social Evolution." *Annals of Human Biology*, vol. 36, no. 5, 2009, pp. 562–572, doi:10.1080/03014460902960289.

Gagneux, Pascal, and Ajit Varki. "Genetic Differences between Humans and Great Apes." *Molecular Phylogenetics and Evolution*, vol. 18, no. 1, 2001, pp. 2–13, doi:10.1006/mpev.2000.0799.

Galand, Pierre E., *et al.* "Diet Shapes Cold-Water Corals Bacterial Communities." *Environmental Microbiology*, vol. 22, no. 1, 2020, pp. 354–368, doi:10.1111/1462-2920.14852.

Gupta, Vinod Kumar, *et al.* "Geography, Ethnicity or Subsistence-Specific Variations in Human Microbiome Composition and Diversity." *Frontiers in Microbiology*, vol. 8, 2017, doi:10.3389/fmicb.2017.01162.

Hammer, M. F., *et al.* "Genetic Evidence for Archaic Admixture in Africa." *Proceedings of the National Academy of Sciences*, vol. 108, no. 37, 2011, pp. 15123–15128, doi:10.1073/pnas.1109300108.

Hanning, Irene, and Sandra Diaz-Sanchez. "The Functionality of the Gastrointestinal Microbiome in Non-Human Animals." *Microbiome*, vol. 3, no. 1, 2015, doi:10.1186/s40168-015-0113-6.

Heijtz, R. D., *et al.* "Normal Gut Microbiota Modulates Brain Development and Behavior." *Proceedings of the National Academy of Sciences*, vol. 108, no. 7, 2011, pp. 3047–3052, doi:10.1073/pnas.1010529108.

Hurst, Gregory D. D. "Extended Genomes: Symbiosis and Evolution." *Interface Focus*, vol. 7, no. 5, 2017, www.ncbi.nlm.nih.gov/pmc/articles/PMC5566813/, doi:10.1098/rsfs.2017.0001.

Kasting, J. F., and Janet L. Siefert "Life and the Evolution of Earth's Atmosphere." *Science*, vol. 296, no. 5570, 2002, pp. 1066–1068, doi:10.1126/science.1071184.

Kurilshikov, Alexander, *et al.* "Host Genetics and Gut Microbiome: Challenges and Perspectives." *Trends in Immunology*, vol. 38, no. 9, 2017, pp. 633–647, doi:10.1016/j.it.2017.06.003.

Ley, Ruth E., *et al.* "Ecological and Evolutionary Forces Shaping Microbial Diversity in the Human Intestine." *Cell*, vol. 124, no. 4, 2006, pp. 837–848, doi:10.1016/j.cell.2006.02.017.

Ley, Ruth E., *et al.* "Worlds Within Worlds: Evolution of the Vertebrate Gut Microbiota." *Nature Reviews Microbiology*, vol. 6, no. 10, 2008a, pp. 776–788, www.ncbi.nlm.nih.gov/pubmed/18794915, doi:10.1038/nrmicro1978.

McFall-Ngai, Margaret, *et al.* "Animals in a Bacterial World, a New Imperative for the Life Sciences." *Proceedings of the National Academy of Sciences*, vol. 110, no. 9, 2013, pp. 3229–3236, doi:10.1073/pnas.1218525110.

Moeller, Andrew H., and Jon G. Sanders. "Roles of the Gut Microbiota in the Adaptive Evolution of Mammalian Species." *Philosophical Transactions of the Royal Society B: Biological Sciences*, vol. 375, no. 1808, 2020, p. 20190597, doi:10.1098/rstb.2019.0597.

Moran, Nancy A., *et al.* "Evolutionary and Ecological Consequences of Gut Microbial Communities." *Annual Review of Ecology, Evolution and Systematics*, vol. 50, no. 1, 2019, pp. 451–475, doi:10.1146/annurev-ecolsys-110617-062453.

Ochman, Howard, *et al.* "Evolutionary Relationships of Wild Hominids Recapitulated by Gut Microbial Communities." *PLoS Biology*, vol. 8, no. 11, 2010, p. e1000546, doi:10.1371/journal.pbio.1000546.

Org, Elin, *et al.* "Genetic and Environmental Control of Host-Gut Microbiota Interactions." *Genome Research*, vol. 25, no. 10, 2015, pp. 1558–1569, doi:10.1101/gr.194118.115.

Patterson, Nick, *et al.* "Genetic Evidence for Complex Speciation of Humans and Chimpanzees." *Nature*, vol. 441, no. 7097, 2006, pp. 1103–1108, www.nature.com/articles/nature04789, doi:10.1038/nature04789.

Pinto-Sanchez, Maria Ines, *et al.* "Probiotic *Bifidobacterium Longum* NCC3001 Reduces Depression Scores and Alters Brain Activity: A Pilot Study in Patients with Irritable Bowel Syndrome." *Gastroenterology*, vol. 153, no. 2, 2017, pp. 448–459.e8, www.ncbi.nlm.nih.gov/pubmed/28483500, doi:10.1053/j.gastro.2017.05.003.

Prescott, Susan L., *et al.* "Biodiversity, the Human Microbiome and Mental Health: Moving toward a New Clinical Ecology for the 21st Century?" *International Journal of Biodiversity*, vol. 2016, 2016, pp. 1–18, doi:10.1155/2016/2718275.

Previc, Fred H. "Dopamine and the Origins of Human Intelligence." *Brain and Cognition*, vol. 41, no. 3, 1999, pp. 299–350, www.sciencedirect.com/science/article/abs/pii/S0278262699911296?via%3Dihub, doi:10.1006/brcg.1999.1129.

Previc, Fred H. "The Role of the Extrapersonal Brain Systems in Religious Activity." *Consciousness and Cognition*, vol. 15, no. 3, 2006, pp. 500–539, doi:10.1016/j.concog.2005.09.009.

Qin, Junjie, *et al.* "A Human Gut Microbial Gene Catalogue Established by Metagenomic Sequencing." *Nature*, vol. 464, no. 7285, 2010, pp. 59–65, doi:10.1038/nature08821.

Roberton, D. M., *et al.* "Milk Antigen Absorption in the Preterm and Term Neonate." *Archives of Disease in Childhood*, vol. 57, no. 5, 1982, pp. 369–372, doi:10.1136/adc.57.5.369.

Sanders, Jon G., *et al.* "Dramatic Differences in Gut Bacterial Densities Correlate with Diet and Habitat in Rainforest Ants." *Integrative and Comparative Biology*, vol. 57, no. 4, 2017, pp. 705–722, doi:10.1093/icb/icx088.

Sender, Ron, *et al.* "Revised Estimates for the Number of Human and Bacteria Cells in the Body." *PLOS Biology*, vol. 14, no. 8, 2016, p. e1002533, doi:10.1371/journal.pbio.1002533.

The Human Microbiome Project Consortium. "Structure, Function and Diversity of the Healthy Human Microbiome." *Nature*, vol. 486, no. 7402, 2012, pp. 207–214, doi:10.1038/nature11234.

Turnbaugh, Peter J., *et al.* "The Human Microbiome Project." *Nature*, vol. 449, no. 7164, 2007, pp. 804–810, www.ncbi.nlm.nih.gov/pmc/articles/PMC3709439/, doi:10.1038/nature06244.

Van den Abbeele, Pieter, *et al.* "The Host Selects Mucosal and Luminal Associations of Coevolved Gut Microorganisms: A Novel Concept." *FEMS Microbiology Reviews*, vol. 35, no. 4, 2011, pp. 681–704, doi:10.1111/j.1574-6976.2011.00270.x.

Varki, Ajit, and Tasha K. Altheide. "Comparing the Human and Chimpanzee Genomes: Searching for Needles in a Haystack." *Genome Research*, vol. 15, no. 12, 2005, pp. 1746–1758, doi:10.1101/gr.3737405.

Vuong, Helen E., *et al.* "The Microbiome and Host Behavior." *Annual Review of Neuroscience*, vol. 40, no. 1, 2017, pp. 21–49, doi:10.1146/annurev-neuro-072116-031347.

Walter, Jens, and Ruth Ley. "The Human Gut Microbiome: Ecology and Recent Evolutionary Changes." *Annual Review of Microbiology*, vol. 65, no. 1, 2011, pp. 411–429, doi:10.1146/annurev-micro-090110-102830.

Weaver, L. T., *et al.* "Intestinal Permeability in the Newborn." *Archives of Disease in Childhood*, vol. 59, no. 3, 1984, pp. 236–241, doi:10.1136/adc.59.3.236.

Winans, Stephen. C., and Bonnie. L. Bassler. "Mob Psychology." *Journal of Bacteriology*, vol. 184, no. 4, 2002, pp. 873–883, doi:10.1128/jb.184.4.873-883.2002.

Xiao, Liang, *et al.* "A Catalog of the Mouse Gut Metagenome." *Nature Biotechnology*, vol. 33, no. 10, 2015, pp. 1103–1108, www.nature.com/articles/nbt.3353, doi:10.1038/nbt.3353.

Zoetendal, Erwin G., *et al.* "A Microbial World within Us." *Molecular Microbiology*, vol. 59, no. 6, 2006, pp. 1639–1650, pubmed.ncbi.nlm.nih.gov/16553872/, doi:10.1111/j.1365-2958.2006.05056.x.

6

HOST MICROBIOME

Living with Commensal Bacteria

A mammal is a complex organism consisting of eukaryotic animal cells and eukaryotic and prokaryotic microbial cells. A large proportion of the microorganisms form communities in the gastrointestinal tract (Savage, 1986). As shown by Gordon and Pesti (1971), the undefined, indigenous intestinal flora and a variety of its known elements can be readily established in germ-free animals. Hence, on experimental grounds, *gnotobiotic* animals (inoculated with specific microorganisms) have gradually become more readily available, and techniques have been simplified to general utility.

Bäckhead et al. (2005) considered that the distal human intestine represents an anaerobic bioreactor programmed with an enormous population of bacteria. The authors further stated that this microbiota and its collective genomes provide humans with genetic and metabolic attributes humans have not evolved on their own, including the ability to harvest otherwise inaccessible nutrients.

Turnbaugh et al. (2009) reported on microbiota plasticity based on experiments in mice – colonised at specific life stages with different microbial communities and referred to as *gnotobiotic* animals,

> Reciprocal transplants involving various combinations of donor and recipient diets revealed that colonisation history influences the initial structure of the microbial community, but that these effects can be rapidly altered by diet. Humanised mice fed the Western diet have increased adiposity; this trait is transmissible via microbiota transplantation.

This observation could be projected onto the impact of unbalanced diets on human infant weight distortions related to socioeconomic conditions.

Microbiota plasticity operates interactively with host enteric dynamics. In this regard, based on the evolutionary context, Lyte (2010) proposed the concept of functional integration of the gut microbiome community and the enteric mammalian nervous system,

DOI: 10.4324/9781032698380-6

Based on the ability of bacteria to both recognise and synthesise neuroendocrine hormones, it is hypothesised that microbes within the intestinal tract comprise a community that interfaces with the mammalian nervous system that innervates the gastrointestinal tract to form a microbial organ … it is further hypothesised that this microbial organ enters into a symbiotic relationship with its mammalian host to influence both homeostasis (aspects such as behaviour) and susceptibility to disease.

According to Norris et al. (2012), more than 3 billion years of evolution have honed the capacities of bacteria to exploit their environments. Their natural history, combined with their extended coevolution with their hosts, would have presumably undergone a selective process favouring those bacteria that best manipulate or interact with their hosts, as discussed in Lyte (2010). Norris et al. (2012) considered that there would be a positive feedback relationship between the microbiome composition of the gut and food preferences and control of host appetites.

*

According to Evans et al. (2013), microbiota constitutes a virtual organ since it lacks *a fully differentiated and functional unit*. However, a long evolutionary history has forged elaborate host-microbial symbioses over several millennia. Microorganisms and their hosts communicate through an array of hormonal signals. This cross-kingdom cell-to-cell signalling involves small molecules, whether hormones produced by eukaryotes or hormone-like chemicals produced by bacteria. Signalling between bacteria, usually called *quorum sensing*, would not be restricted to bacterial communication but also allows communication between microorganisms and their hosts (Hughes and Sperandio 2008).

Clarke et al. (2014) supported the concept that gut microbiota represents an endocrine organ due to its ability to influence the function of distal organs and systems by releasing *hormone-like* products into the interstitial tissue. They are conveyed by blood and lymph capillaries to act remotely, usually at low concentrations, including the brain. Nevertheless, as stated by these authors, unlike other endocrine organs, the microbiota has a remarkable plasticity and can rapidly and significantly alter its composition and molecular interactions in response to diet and host physiological state. In this regard, as mentioned by Garud et al. (2019), the gut microbiota is shaped by a combination of ecological and evolutionary forces. According to Lopez-Goñi (2018), humans are *superorganisms* in which 1% of their genome is inherited from the parents and 99% from microbes. Regarding the collection of microorganisms living in a host (the *microbiome*), as stated by Henry et al. (2021),

Microbiomes are traditionally viewed as nongenetic, environmental factors that influence host phenotypes. However, unlike abiotic environmental conditions, the effects of microbial variation have a genetic basis and can evolve10 but are not always inherited in the same way as host genes. While microbiomes have substantial phenotypic effects on their hosts, these effects strongly depend on the ecological context.

As a result of the close symbiotic host association with the microbiome in all animal species, for our purpose, most specifically in humans, the result could be considered as superorganisms involving variable host (genomic) indigenous metabolic processes and those of the microbiome. As posed by Li et al. (2008), superorganism metabolism involves integrating truly indigenous metabolic processes (coded in the host genome) with those of the microbiome. This concept represents a significant paradigm shift in understanding human biology and its impact on diverse metabolic deviations. Genetically homogeneous animals can have diverse metabolic phenotypes when they have structurally different gut microbiota. As commented by Jansma and Aidy (2021), the metabolic output of microorganisms depends on their surroundings, which conditions the metabolic output and host-microbiome interactions on individual interactive profiles.

Hence, several authors have proposed the term superorganisms for this host-microbiome collective, a concept though questioned by Foster et al. (2017), who considered that we cannot assume that the host and microbiota are a single evolutionary unit acting with a common interest, as is sometimes done in applications of the "holobiont" or "superorganism" metaphors. This, given that the host and each microbial strain are distinct entities with potentially divergent selective pressures. Though this conceptual disagreement remains unsettled, the holobiont concept has received support, as discussed throughout this text.

*

Mira et al. (2006) placed a comparative perspective associated with the so-called Neolithic revolution, with an increase in bacterial transposable elements, indicators of specialisation within human-related niches. This is associated with large and stable human communities, agriculture and animal domestication: three features unequivocally linked to the Neolithic revolution. In other words, the authors stressed that bacteria specialised in human-associated niches underwent an intense transformation after the social and demographic changes with the first Neolithic settlements. During this period, specifically during the Mesolithic-Neolithic transition, as stated by Bocquet-Appel (2002), the advent of agriculture and animal husbandry brought the most significant social revolution in the history of humankind and its microbiome association.

Climate-driven changes in habitats are thought to have favoured significant alterations in the foraging behaviour and diet of early *Homo* species. Coupled with animal evolution, whether in symbiosis or via shared habitats, it has also influenced the distribution and diversification of bacteria. Thus, the evolution of animals provided novel physical environments for bacterial colonisation. Hence, as early animals diversified, animal–bacterial interactions continued to shape evolution in new ways (McFall-Ngaia et al. 2013). According to Nishida and Ochman (2018), the number of dietary transitions within a lineage does not influence rates of microbiome divergence. However, some of the most dramatic changes are associated with the loss and renewal of bacterial taxa, such as those accompanying the transition from terrestrial to marine lifestyles and the evolution of hominids.

Regarding the impact of shifting from hunter-gatherers to sedentary profiles, Walter and Ley (2011) considered that,

The development of agriculture and the domestication of animals have been major factors in recent human evolution. Very early hominids were largely omnivorous and subsisted during periods of food scarcity on starch-rich roots and bulbs, especially in savannahs where edible plants were scarce. In the past 13,000 years, food production arose independently in several areas of the world, and diets were broadened to include high-starch plant foods and dairy products. We suggest that the subsequent changes in diet set the stage for intense conflict with diet substrates between host and gut microbiota.

Lombardo (2007) posed a central question in behavioural ecology regarding cell/animal grouping. This would be primarily associated with the evolution of sociality focused on the potential benefits of decreased risk of predation, increased foraging or feeding efficiency, and mutual aid in defending resources and/or rearing offspring. The author's argument is based on access to mutualistic endosymbiotic microbes as an underappreciated evolutive benefit of group living.

Nishida and Ochman (2018) studied microbiome data from a phylogenetically diverse cohort of over 100 mammalian host species to analyse the pace of evolution in microbial community composition and the factors that contribute to gut microbiome divergence. Despite the potential for significant shifts in the microbiome composition of individuals and the diversity within populations due to diet, mammalian species could be distinguished based on their microbiome compositions. Among primates, which over their 75-million-year history have sustained the most constant rate of microbiome divergence among mammals, there has been an acceleration in the human lineage due to the dramatic loss of bacteria at all taxonomic levels, as described in Moeller et al. (2014). Nishida and Ochman (2018) considered that despite the high levels of social behaviour in humans, the reduction of microbial diversity and accelerated divergence of human microbiomes have been ascribed to changes in lifestyle, including dietary features, improved sanitation, and use of antibiotics.

On species comparative grounds, rates of microbiome divergence have significantly accelerated in *Cetartiodactyls* (*Artiodactyla* plus *Cetacea*), accompanying the transition from terrestrial to marine environments, and in Hominids, during the evolution of the hominid lineage (Nishida and Ochman, 2018). The number of dietary transitions within a lineage does not influence rates of microbiome divergence, but instead, some of the most dramatic changes are associated with the loss of bacterial taxa. In 2019, Nishida and Ochman provided an evolutionary perspective exposing how human gut microbiomes have been shaped by our great-ape heritage and the features that make humans unique.

*

Each human is an assemblage composed not only of somatic cells but also of many symbiotic species. The abundant and diverse microbial members of the assemblage play critical roles in the maintenance of human health by liberating nutrients and/or energy from otherwise inaccessible dietary substrates, promoting differentiation of host tissues, stimulating the immune system and protecting the host from invasion by pathogens.

(Costello et al. 2012)

As summarised by Rothschild et al. (2018), the gut microbiome is increasingly recognised as having fundamental roles in human physiology. Further, the authors considered that increasing evidence suggests that microbiome composition is dominated by local environmental factors rather than host genetics.

According to Ley et al. (2006),

> ... the human gut is populated with as many as 100 trillion cells, whose collective genome, the microbiome, is a reflection of evolutionary selection pressures acting at the level of the host and at the level of the microbial cell. The ecological rules that govern the shape of microbial diversity in the gut apply to mutualists and pathogens alike.

This general statement regarding the human case with its cultural diet variations, as it will be discussed later, must be expanded to include comparative variations that each animal species represents according to its environmental niche and feeding habits.

As mentioned by Coyte et al. (2015), the human microbiome provides health benefits, such as the breakdown of complex molecules in food, protection from pathogens, and healthy immune development. Though highly individualised, any individual tends to carry a given set of species (symbionts) for long periods in a rather stable microbiome ecological community, thus ensuring relative stability on its beneficial effect on host health. Gut microbiota communities are assembled each generation and depend on maternal seeding, environmental factors, host genetics, and age, which impose individual variations in metabolic traits among human populations (Org et al. 2015).

According to Costello et al. (2012), health is a collective property of the human body and its associated microbiome and thus could be considered a net effect of ecosystem services. Regarding its impact on the offspring's health, as Pronovost and Hsiao (2019) discussed, the maternal and neonatal microbiome is increasingly recognised as an important regulator of health and predisposition to later-life diseases. As these authors posed, much of the evidence supporting the early life microbiome's causal effects on disease symptoms derives from animal models that utilise reductionist approaches. Hence, though some behavioural pathological outcomes are known to be produced by viral and bacterial presence in different species, microbiome impact on mental/cognitive performance is challenging. This is compounded by the multiple gut microbe commensalisms.

*

Let us consider the previous mentions of the ancestral origin of simple living forms, their ubiquity and their capacity to dwell and reproduce in harsh environments as they were in primaeval times, perhaps less than 4 million years ago. Still more, let us include their endurance in extremely harsh environments for some forms, as commented earlier. If Natural Kingdom evolution represents a blind chain of adaptive, survival-prone events that in the distant past gave origin to its three main ubiquitous groups (Archaea, Bacteria, Eukaryote), one odd and unsettling question arises. Under those conditions, evolution would represent an unimaginable (considering the species extinction processes) series of frames mounting a metaphoric sequence of how original cell forms and their simple organismic derivates managed to survive and colonise Earth, generating an incredible rainbow of vegetable and animal species, temporarily and successively adapted to different ecosystems. The evolution of the alimentary tract of metazoan provided new environments for microbial thrive, unlike those previously available ecologic niches. As described in previous paragraphs, microbes suited for these niches exerted significant survival pressures on the evolution of vertebrates.

> The long history of shared ancestry and alliances between animals and microbes is reflected in their genomes. Analysis of the large number of full genome sequences presently available reveals that most life forms share approximately one third of their genes, including those encoding central metabolic pathways. Not surprisingly, many animal genes are homologs of bacterial genes, mostly derived by descent, but occasionally by gene transfer from bacteria. For example, 37% of the 23,000 human genes have homologs in the Bacteria and Archaea, and another 28% originated in unicellular eukaryotes. Among these homologous genes are some whose products provide the foundation for signalling between extant animals and bacteria.
>
> *(McFall-Ngaia et al. 2013)*

Under this perception, humans would represent another link contributing to expanding microbial domains, assuring an efficient means to invade other environmental dimensions, whether terrestrial or extra-terrestrial. Hidden in human species pride, our own, as well as other living species, would represent a means for a blind, opportunistic spread of ancient microbiota. Mounted on these evolutive dynamics, though, evolution has provided grounds for the emergence of a species able to inquire, evolve, and adopt natural and technological creative survival strategies through expanding cumulative knowledge. A new dominant species (Colombo 2022) yet construed on primal host-microbiome interactions.

As commented by Stilling et al. (2015),

> … animals have never lived in a sterile environment and will never be able to live, develop nor evolve in a germ-free way outside laboratory isolators. Best documented by the existence of mitochondria and chloroplasts, microbes became part of multicellular life forms right from the beginning and today share a long history of co-adaptation and mutual influence on evolutionary trajectories. Given this intimate relationship, it seems inappropriate to imply conflicting interests for microbiota and host.

As Goulet (2015) mentioned, the gut microbiota protects against pathogens by competing for nutrients and receptors, producing antimicrobial compounds, and stimulating a multiple-cell signalling process that can limit the release of virulence factors. The gut microbiota also influences the development of the intestinal barrier and its functions, which could be affected by factors that may influence early intestinal colonisation (prematurity, caesarean section, breastfeeding, antibiotics) (see also Arrieta et al. 2015) and its final maturation, such as early malnutrition. Additionally, Johnson (2019) noted that differences in gut microbiome composition and diversity are shown to be linked to personality traits in the general population. This adds a new dimension to our understanding of personality, in line with accumulating evidence that the gut microbiome can influence a human's central nervous system, affecting behaviour. This consideration appears in line with observations made in humans by Kurokawa et al. (2018) following the analysis of the effect of faecal microbiota transplantation on psychiatric symptoms among patients with gastrointestinal disease.

*

The Human Context: Microbiome Inoculum in Newborns, Host Diet

The same Escherichia coli serotypes were found in both the mouths of babies immediately after birth and in their mothers' feces, implying that during natural birth microbes from mothers' feces contaminate infants. The gastric content of 5–10-min-old babies was similar to that of their mothers' cervix. Also, immediately after birth, the nasopharynxes of 62% of babies contained bacteria that were consistent with those of their mothers' vaginas immediately before delivery.

(Mackie et al. 1999)

Gut microbes are generally acquired after birth or hatching, from conspecific hosts and/or from other environmental sources, depending on the animal species. Gregarious or social animals have more opportunities for direct acquisition from conspecifics than do solitary species. In turn, frequent direct transmission provides greater opportunity for gut bacteria to specialise on the gut niche and to lose the ability to replicate outside of the gut environment.

(Moran et al. 2019)

The microbiota comprises bacteria, viruses, fungi, and other microorganisms living mutualistically in and on animals. The universe of microbes colonising the human digestive tract would include an estimated number of 150 to 400 species, with their number and relative proportion varying according to several conditioning variables, such as individual, age, ethnicity, and feeding habits. According to Qin et al. (2010), 99.1% of human gut microbiome catalogue genes are of bacterial origin, the remainder primarily archaeal, with only 0.1% of eukaryotic and viral origins. Regarding the Archaea, Kim et al. (2020) commented that they were separated from prokaryotes based on ribosomal RNA gene sequences and mostly considered extremophiles. However, mesophilic Archaea have been identified in several substrates in moderate environments. The authors have observed a relative abundance of Archaea (10.24 ± 4.58% of the total bacterial and archaeal abundance) in the human gut. Regarding heritability, as quoted in Henry et al. (2021), not all microbiome components are heritable, with estimates ranging from 8–56% of microbes being transmitted faithfully.

Moeller et al. (2014) considered that primates, including humans, are ecosystems containing trillions of microorganisms influenced by their lifestyle – a statement that perhaps should be expanded to most living macro-species. The authors based their analysis on sequencing ribosomal DNA in faecal samples from several sources. These included wild chimpanzees, bonobos, gorillas, and human sources from the urban USA, rural lifestyles in Malawi, preindustrial in the southern Amazon rainforests of Venezuela, urban lifestyles in Europe, and hunter-gatherer lifestyles in Tanzania. According to the authors, humanity has experienced a depletion of the gut flora since diverging from *Pan*. This is represented by human microbiomes losing ancestral microbial diversity while becoming specialised for animal-based diets.

As mentioned by Whiten and Erdal (2012), on a human evolutive dimension, dietary successful hunter-gatherer foraging provided the essential foundation for the rest of early human tribes. According to these authors, this event would be associated with the evolution of a new socio-cognitive niche that involves forms of cooperation, egalitarianism, mindreading ("theory of mind"), language, and cultural transmission that go far beyond the most comparable phenomena in other primates. These cognitive and behavioural aspects allowed hunter-gatherer bands to function as a unique and highly competitive predatory organism, impinged upon feeding characteristics and host microbiome.

Moeller and Sanders (2020) stated that the gut microbiome benefits mammalian species by enabling host transitions to novel dietary niches and subsequent diversification. Thus, amplifying environmental signals important for host fitness would have contributed to the canalisation of host developmental processes, providing an expanded metabolic potential. Furthermore, these authors affirmed that,

> … natural history studies have indicated that most mammalian species harbour compositionally distinct gut microbiotas that reflect the phylogenetic histories of their hosts. Moreover, experimental studies have shown that gut microbiotas are in many cases deeply intertwined with mammalian phenotypes spanning neuroendocrine, immune and metabolic systems, including traits related to fitness and adaptive differences between species.
>
> Our understanding of species evolution is undergoing restructuring. It is well accepted that host–symbiont coevolution is responsible for fundamental aspects of biology. However, the emerging importance of plant and animal associated microbiotas to their hosts suggests a scale of coevolutionary interactions many-fold greater than previously considered.
>
> *(Shapira 2016)*

The molecular basis of microbe-host interactions and the roles of individual bacterial species require further insight. It is generally accepted that, as stated in Ley et al. (2006), the microbial diversity (microbiome) of the human gut is the result of coevolution between microbial *communities,* a consortium of gut microbes and their hosts with which have evolved an intimate symbiotic relationship. As stated by these authors, defining the gut microbiota and, in general, the microbiome in people living in various geographic regions and under different levels of economic development should provide an opportunity to monitor human "microevolution", more yet, since humans traverse a period of profound social, economic, and ecological changes. Added to these variables, Blekhman et al. (2015) stressed the role of host genetic variation in shaping the composition of the human microbiome.

As reported by Penders et al. (2006), the most important determinants of the gut microbiome composition in human infants are the mode of delivery, type of infant feeding, gestational age, and infant hospitalisation. According to these authors, term infants born vaginally at home and breastfed exclusively seemed to have the most beneficial gut microbiota (highest numbers of *bifidobacteria* and lowest numbers of *C difficile* and *E coli*). Kurokawa et al. (2007) – in what would constitute the first large-scale comparative metagenomic analysis of human gut microbiomes – reported that while the gut microbiota from non-weaned infants was simple and showed a high inter-individual variation in taxonomic and gene composition, those from adults and weaned children were more complex but showed a high functional uniformity regardless of age or sex. The authors considered that the infant-type can be viewed as unstable yet dynamic and adaptable, while the functional uniformity observed in the adult-type microbiota may be attributable to its more complex nature. Amato (2016) considered the impact of intimate physical interactions of the mother with her offspring and quoted evidence that infants are inoculated by their mothers' vaginal microbiota and continue to acquire microbes after birth through contact with conspecifics and the environment. Addressing experimental grounds, the author considered that gut microbiota composition in mice fed the same diet was more similar between mother and offspring than between unrelated individuals. Given these maternal effects, an element of vertical transmission would exist for the mammalian gut microbiota. In addition, the authors supported the concept that the gut microbiota would likely play a role in the development of cognition during childhood. In fact, according to Funkhouser and Bordenstein (2013), maternal microbial transmission has been reported to occur in all animal kingdoms.

The developing gut microbiome would undergo a progressive increment during early postnatal life, as mentioned by Stewart et al. (2018). Breast milk intake was the most significant factor associated with the microbiome structure during this developmental period. Microbiome stabilisation, in which infants' samples remained in the same cluster at consecutive time points, was observed from the 31st month of life. According to Stinson (2020), the microbiome undergoes a highly dynamic growth phase during the first three years of life, during which it is susceptible to maldevelopment and can cause later-life disease. Coyte et al. (2021) considered that microbe-to-microbe and host-to-microbe interactions are crucial for the microbiome assembly, which occurs gradually after birth in several species.

Regarding microbiome developmental inoculation, Funkhouser and Bordenstein's (2013) studies suggest that infants incorporate initial microbiomes before birth and receive copious maternal microbiome inoculum through birth and breastfeeding. Hence, it would be possible to identify entire microbiomes that are transferred from mother to offspring in humans. Coincidentally, as stated by Aagaard (2014), collective observations raised the possibility that the infant may be first seeded in utero by a low abundance source from the placenta, which may vary by gestation length. These authors concluded that the placenta harbours a low abundance but metabolically rich microbiome. This established early colonisation would be seeded by haematogenous oral microbiota spread during early vascularisation and placentation. According to Funkhouser and Bordenstein (2013), maternal microbial transmission has been reported to occur in all animal kingdoms.

In such regard, the concept of sole postpartum colonisation has been challenged by observation of prenatal colonisation, as stated by Gilbert (2014) and Rodriguez et al.

(2015). According to these authors, there is increasing evidence of early microbial contact, and it was suggested that human intestinal microbiota is seeded before birth, further stating that recent findings challenge the dogma that the foetus resides in a sterile environment. These authors also considered the recent extensive, profound sequencing studies in healthy term pregnancies, identifying a low abundance but metabolically rich placental microbiome. Its composition would resemble the oral microbiome more than the vaginal, faecal, skin, or nasal microbiomes. As also quoted in Dinan et al. (2015), there is an increasing body of evidence challenging the sterile womb paradigm and that transmission of certain microbes already occurs in utero, as stated in the review by Clapp et al. (2017). This concept was supported by Gensollen et al. (2016), who considered that,

> ... the exposure of the mammal to microbiota begins in utero and expands rapidly after birth. For example, maternal gut bacteria can be detected in the amniotic fluid of pregnant mice, and bacteria can be isolated in the meconium from preterm human babies.

This view is also shared by Dominguez-Bello et al. (2019), who considered it possible that some bacterial cells of the uterine cervix may enter with the sperm during fertilisation and reach the egg at the time of fertilisation, implantation, or early embryonic development.

However, this view of the first inoculation raised controversy. Moore and Townsend (2019) cast doubts on the methods that support previous conclusions and concluded that based on the available data, a human being's first microbial inoculation would occur during labour and delivery. Additionally, Perez-Muñoz et al. (2017) argued against the "in utero colonisation hypothesis," stating that current scientific evidence does not support the existence of microbiomes within the healthy foetal milieu.

Passage through the birth canal exposes the baby to the mother's microbiota, therefore initial colonisation is dictated by the mother's microbes and the hospital environment.

Adding to the discussion regarding microbial gut colonisation and early establishment of the microbiome, McBurney et al. (2019) considered that,

> Microbial colonisation of the human body begins postpartum and proceeds in an incremental manner from infancy to adulthood, with the largest microbial community being found in the distal regions of the adult human gastrointestinal tract. Intestinal colonisation occurs during infanthood and is affected by mode of delivery, diet, probiotic supplementation, antibiotic use, and possibly maternal microbiome during pregnancy.

Infants are naturally born with their skin and mouth covered by maternal inoculum and having swallowed these microbes. According to these authors, we inherit the primordial microbiota from our mothers, grandmothers and further on the matrilineal line, with microbial vertical transmission extending back to earlier ancestors.

In a comparative study, Moeller et al. (2014) observed that,

Relative to the microbiomes of wild apes, human microbiomes have lost ancestral microbial diversity while becoming specialised for animal-based diets. Individual wild apes cultivate more phyla, classes, orders, families, genera, and species of bacteria than do individual humans across a range of societies. These results indicate that humanity has experienced a depletion of the gut flora since diverging from Pan.

Additionally,

> ... multicellular eukaryotes do not take up exogenous DNA as readily as microbes: instead, they form symbiotic associations with microbes that carry the necessary genes, allowing a rapid adaptive extension of their phenotypic capabilities [78]. Host microbe symbiosis is widely distributed within the Eukaryota ...
>
> *(Walter and Ley 2011)*

Stewart et al. (2018), based on stool samples from 903 children, concluded that there are three stages of gut microbiome development: the developmental phase (3–14 months), the transitional phase (15–30 months) and the stable phase (31–46 months) (see also, Koenig et al. 2011). Regarding the determinants of infant microbiome diversity, in addition to maternal inoculum, Moore and Townsend (2019) commented that,

> ... mode of delivery, breastfeeding versus formula feeding, antibiotic use and introduction of solid foods, environmental exposures can also play a key role in the variability of that microbiota. Hospital setting, cohabitation with family members, geographical location, air quality, pet and animal exposure, and daycare are all included in the environment factors that contribute to neonatal microbiome development.

McBurney et al. (2019) considered that in this host-microbiome universe,

> Microbial colonisation of the human body begins postpartum and proceeds in an incremental manner from infancy to adulthood, with the largest microbial community being found in the distal regions of the adult human gastrointestinal tract. Intestinal colonisation occurs during infanthood and is affected by mode of delivery, diet, probiotic supplementation, antibiotic use, and possibly maternal microbiome during pregnancy.

The analysis undertaken by Valles-Colomer et al. (2022) would confirm the existence of family-bound microbiome community profiles. According to the authors, this suggests that transmission or co-acquisition of bacterial strains would be linked to cohabitation.

According to Ferretti et al. (2018), strain-level metagenomic profiling showed a rapid influx of microbes at birth followed by strong selection during the first few days of life and suggests that intra-uterine seeding is still debatable. This view is shared by Gaufin et al. (2018), who noted that the reported detection of microbes likely represents background contamination of the reagents, which plagues the analysis of all low biomass studies.

To date, most infants would have no bacteria at birth, though the definitive answer to this question remains unsettled. The concept of prenatal microbial seeding is also confronted by Dalby and Hall (2020), who stated that,

> The womb has traditionally been considered largely sterile; however, some previous studies detected microbial signatures after DNA sequencing of placenta, amniotic fluid, and meconium samples. If indeed present, the low DNA yields indicate that any bacteria in the womb would be in very low numbers, with the various genus of bacteria identified not appearing to colonise the infant after birth. Thus, the effect of such bacterial exposure before birth is unlikely to form a key pathway for seeding of the neonatal gut microbiota.

These authors further considered that the importance of the milk microbiota remains to be explored (Moosavi and Azad 2019). Hence, the establishment of the gut microbiota in infants is an ecological succession (an issue that would still be unsettled regarding early events) shaped by sources of exposure to different microbes over time, which can be potentially disrupted by exogenous factors (see also Bogaert et al. 2023).

Dominguez-Bello et al. (2010) in their study based in Puerto Ayacucho Hospital, Amazonas State, Venezuela, observed that the newborns harboured bacterial communities that were essentially undifferentiated across skin, oral, nasopharyngeal, and gut habitats regardless of delivery mode (vaginal or caesarean) showing that in its earliest stage of community development, the human microbiota is homogeneously distributed across the body (see also Backhed et al. 2015).

Sharon et al. (2016) considered that the first direct encounter an infant has with the microbial world would be during birth, and this depends on the delivery conditions, whether through routine procedures or under a caesarean intervention. In the latter case, during the initial postnatal exposure, infants would be colonised by skin microbes that, according to Sharon et al. (2016), would have long-term health and developmental consequences. Furthermore, on pathological grounds, according to Sharon et al. (2019), transplantation of gut microbiota from human donors with *autism spectrum disorder* (ASD) into germ-free mice revealed that colonisation with ASD microbiota is sufficient to induce hallmark autistic behaviours. Thus, according to the former authors, microbiome and metabolome profiles of mice harbouring human microbiota predict that specific bacterial taxa and their metabolites modulate ASD behaviours.

*

Besides delivery mode, according to Zijlmans et al. (2015), infant gut microbiota composition is affected by the mother's prenatal stress, whether based on reported stress, elevated basal maternal salivary cortisol concentrations, or both. Furthermore, the colonisation composition pattern was related to more maternally reported infant gastrointestinal symptoms and allergic reactions. Similar results were obtained by Aatsinki et al. (2020) based on a target population from the Finn Brain Birth Control

Study. Hechler et al. (2019) considered that maternal prenatal psychosocial stress is associated with altered child emotional and behavioural development and that one potential underlying mechanism is that prenatal psychosocial stress affects child outcomes via the mother's intestinal microbiota. The authors examined associations between maternal psychosocial stress and intestinal microbiota composition in late pregnancy and found associations between maternal general anxiety and microbial composition. These results would provide evidence of how psychological symptoms during pregnancy may affect the offspring.

Additionally, the impact of prenatal stress on gut microbial colonisation was experimentally studied in pregnant Rhesus monkeys (Bailey et al. 2004). It involved individuals undisturbed or subjected to moderate acoustic stress for six weeks at different gestational periods. Developmental changes in intestinal bacteria occurred during the first six months of life in both control and prenatally stressed offspring who expressed altered microflora. The latter could have also involved changes in maternal milk bio flora, which was not tested.

Collectively, these observations would tend to attribute to maternal inoculum, environment, and genetics – albeit with significantly different weight – a crucial role in establishing the gut microbiome. The composition of the vertebrate gut microbiota is further influenced by diet, host morphology, and phylogeny, and in this respect, the human gut bacterial community is typical for an omnivorous primate.

*

Evolutive, Comparative Considerations

The human gut is populated with as many as 100 trillion cells, whose collective genome, the microbiome, reflects evolutionary selection pressures at the host level and the microbial cell (Ley et al. 2006). The microbiome has shaped phenotypes in our ancestral lineages by coevolving with the host. As stated in Ley et al. (2008b), multicellular eukaryotes would have existed for at least 1.2 billion years. Thus, the evolution of the vertebrates was preceded and was probably shaped by a long history of interaction between multicellular life forms and microbial communities. Furthermore, according to these authors,

> Host responses to microbial colonisation are evolutionarily conserved among diverse vertebrates including zebrafish, mice and humans. The scripts that dictate our interactions with our microbial partners thus provide some of the foundations of our Homo sapiens genome.

Stilling et al. (2014) stated that the tight association of the human body with trillions of colonising microbes that we observe today is the result of a long evolutionary history. Therefore, the hologenome concept embraces the contemporary gene-centric view of life but upgrades it to include the microbiome as a central facet of an organism's genetics.

The reported congruence of the phylogenetic trees of intestinal bacterial microbiota and primates demonstrates host-microbiota coevolution and implies within-species transmission of microbes across generations, as stated by Dominguez-Bello et al. (2019), adding that,

> … we inherit the primordial microbiota from our mothers, grandmothers and further on the matrilineal line, with microbial vertical transmission extending back to earlier ancestors. Whether the primordial inoculum contains most microbes that will be nurtured by the child, and which maternal strains colonise which parts of the baby's body and their functions, are still not completely understood.

Qian and Akcay (2020) presented a model with saturating benefits from mutualisms and sequentially assembled communities and showed that such communities are internally stable for any level of diversity and any combination of species interaction types. Furthermore, on the authors' model, a higher fraction of mutualistic interactions can increase the external stability and diversity of communities and species persistence, should its interactions provide unique benefits. According to these authors, ecological selection increases the prevalence of mutualisms, and limits on biodiversity emerge from species interactions.

Contijoch et al. (2019) collected faecal material from 16 mammalian species and observed significant differences in microbiota density among them. Animals from the Order *Carnivora* (dog, ferret, lion, red panda, and tiger) had significantly reduced microbiota densities compared with the other mammals studied. Further, according to Moran et al. (2019), while some animals (humans and termites) host vast numbers of specific microbes that play critical roles in their host's growth and survival, other animals (caterpillars, many though not all ants and stick insects) have few or no resident gut microbes.

As discussed in Evans et al. (2013), the microbiota colonising the gut is considered a virtual organ or emergent system. However, it does not conform to the current organ definition involving a fully differentiated and functional unit (albeit this issue would require further discussion), given that its composition depends on diet and other environmental variables. As the quoted authors state, its properties must be integrated into our understanding of host biology, physiology, and behaviour.

*

According to Douglas (2014), there is now persuasive evidence that extant eukaryotes are derived from an association with intracellular bacteria within the *Rickettsiales* that evolved into mitochondria (cf., Williams et al. 2007), with the implication that this propensity to form persistent associations has very ancient evolutionary roots. According to Sarkar et al. (2020), multicellular life hosts microbial life, and the relationships between microorganisms and host lineages appear stable over millions of years of host evolution. Microbes in animals reside on the skin and mainly in the intestinal tract, where they colonise during parturition. The initial microbial priming is followed by reorganisation based on development, diet, health, and environmental interaction.

Microbial colonisation affects several functions related to energy balance, immunity, and the more recently studied effects on brain development and function, as discussed in Heijtz et al. 2011; Sampson and Mazmanian 2015; Sharon et al. 2016, 2019; Amato 2016; Parashar and Udayabanu 2016; Vuong et al. 2017; and Cussotto et al. 2018. These relationships with brain and social behaviour will be considered under the general scope of factors influencing the construction of human behaviour (including emotion, learning, and memory) under different ecological and cultural conditions. They constitute evidence that these interactions are not foreign to expressing complex human behaviours under the most diverse ecological circumstances.

In this regard, it should be considered that the brain is a costly organ for the economy of our body (Peters et al., 2004), demanding specific nutrients, most specifically during the process of brain and mental development. In terms of human energy requirements, the body requires at birth 60–80% of what it requires in basal conditions. During adult life, the brain demands between 20–25% of total energy consumption under basal conditions (Leonard et al. 2003; Leonard et al. 2007; cf. Colombo 2019), which is modified under stressful conditions (Hitze et al. 2010). However, the adult brain represents about 2% of an individual's body weight. In comparison, a dog's brain consumes about 5% of its total body demand.

Several hypotheses have been proposed to understand the brain and mind development of *Homo sapiens*. Among them, social complexity as an ecological factor (Dunbar 2009; Dunbar and Shultz 2007); shift in feeding behaviour with the subsequent reduction of the digestive system to compensate for energy demand due to brain growth (Aiello and Wheeler 1995); genetic modifications (see before); neurotransmitter involvement, such as dopamine (Previc 1999, 2009), and culture (see ahead). These proposals do not seem to contradict but complement each other, acting at similar or different (sequential) times.

On comparative grounds, Ochman et al. (2010) concluded that evolutionary changes in host physiology occurred among primates during the divergence of great apes. This would have been the dominant factor in shaping each host species' distal gut microbial community, revealing a clear species-specific signature of microbial community structure. Furthermore, the authors stated that over evolutionary timescales, host phylogeny is the overriding factor determining the great ape distal gut microbiota's microbial composition, not diet. In this domain, Moeller et al. (2014) stated that according to their comparative studies among primate orders, humanity has experienced a depletion of the gut flora since diverging from *Pan*. This suggests that the human microbiome has substantially transformed since the human-chimpanzee split, thus projecting a sociocultural impact on human host-gut microbiome interactions.

*

The question of whether host microbiome could affect host behaviour has also been considered on comparative, non-mammalian grounds. As Sampson and Mazmanian (2015) commented, several examples disclose microbial species' impact on host behaviour for its developmental and proliferation goals. In this regard, a profit attraction relationship was also found among flowering plants, as Steiger et al. (2010) mentioned. Sampson and Mazmanian (2015) quoted references involving rodents infected with *Toxoplasma gondii*, leading the host to become attracted to felines where

Toxoplasma continues its life cycle, as House et al. (2011) described. In addition, it affects the water temperature preference of the stickleback fish, where the infesting parasite could increase its growth rate (Macnab and Barber 2011).

Within this domain, the impact of the intestinal microbiome on human regulatory systems and behaviour represents a significant quest in this interactive domain between host and commensal guest. This host-commensal or prey relationship represents perhaps the most ancient interaction among species affecting different goals, whether survival, reproductive, or of a dominance nature. As mentioned by Steiger et al. (2010), behaviours and traits have evolved to influence other organisms, whether as commensal or exteroceptive interactions. Moreover, this could take place in different bioecological domains.

> … organisms, from plants to mammals, release waste products and other chemicals that incidentally carry information. We have argued that such products provide multiple starting points for the evolution of chemical communication thereby explaining the prevalence of this mode of communication and its diversity.
>
> *(Steiger et al. 2010)*

*

Microbiota and Feeding Cultural Profiles

A fundamental question that generated opposing views is the extent to which microbiome composition is determined by host genetics instead of being shaped by environmental variables, such as diet. According to Blekhman et al. (2015) and Goodrich et al. (2014, though see Goodrich 2016), these variations would be affected mainly by genetic influences, highlighting the role of host genetics. Coincidentally, Blekhman et al. (2015) stressed the role of host genetic variation in shaping the composition of the human microbiome. This confronts the view that environmental factors such as diet dominate over host genetics in shaping human gut microbiome composition, as posed by Rothschild et al. (2018),

> In contrast to the lack of association between host genetics and gut microbiome, we found significant correlations between the functional composition of gut microbiomes among individuals sharing the same household. This result corroborates previous studies showing that the human oral microbiome is dominated by household sharing[60], and that diet reproducibly alters the gut microbiota of mice with diverse genotypes[61]. Thus, an increasing body of evidence suggests that microbiome composition is dominated by environmental factors rather than by host genetics.

Based on twin studies, these authors estimated that the overall microbiome heritability lies between 1.9% and 8.1%.

According to Goodrich et al. (2016), the environment would be a lower driver than other complex genetic traits measured in the same population, such as diet-sensing, metabolism, and immune defence.

As will be further discussed, this issue is explored by neurobiological studies that analyse the impact of gene-microbiome environmental interactions.

As stated by Henry et al. (2021), the microbiome extends host evolutionary potential. Integrating microbial genetic variation into host evolutionary processes builds on the "hologenome" theory. In this regard, Ochman et al. (2010) underline the role of host phylogeny in shaping the gut microbiota over evolutionary timescales. Their analysis involving several primate species (*Homo sapiens, Pan troglodytes, P. paniscus, Gorilla gorilla*, and *G. beringei*) revealed a clear species-specific signature of microbial community structure, concluding that host phylogeny is the overriding factor determining the microbial composition of the great ape gut microbiota, rather than diet. Additionally, these authors stated that the gut is initially and continuously seeded by bacteria acquired from external sources. Over evolutionary timescales, the composition of the gut microbiota among great ape species is phylogenetically conserved and has diverged in a manner consistent with vertical inheritance.

Once again, dynamic interactions between host phylogeny and diet are the source of a debatable priority, though combined proposals were included. In this regard, Ley et al. (2008b) proposed an integrated influence on gut microbiome, supporting the concept that host diet and phylogeny both influence bacterial diversity – though assigning a critical role on the diet – which would increase from carnivory to omnivore to herbivory species. This concept was supported by Trevelline et al. (2022), who stated that the microbiome can influence host diet selection behaviour by mediating the availability of essential amino acids, a possible mechanism by which the gut microbiota can influence host foraging behaviour. In this comparative domain, the gut microbiota of humans living a modern lifestyle would be typical of omnivorous primates. Ley et al. (2008a) also observed that,

> Despite large variability between human fecal bacterial communities from healthy men and women from three continents, spanning 27 to 94 years of age, the human-associated communities were more similar to one another than to those associated with members of other mammalian species.

Though adding that,

> … human fecal microbiotas were more similar to those of other primates than to non-primates, but not to other hominids specifically. Instead, diet appeared to be of principal importance in clustering among primates. Human samples clustered with those of other omnivores (e.g., Ring-tailed Lemur, Black Lemur, Mongoose Lemur, Bonobo, Spider Monkey), but the other hominids, which tend have a diet that is more dominated by plant materials, clustered in an intermediate position between the omnivorous primates and non-primate herbivores …

At the human level, Rothschild et al. (2018) posed that family relatives with no history of a shared household do not have similar microbiomes. Microbiome similarity was found among genetically unrelated individuals who share a household, thus supporting the concept that environmental factors have a substantially stronger effect on microbiome composition than host genetics. After screening 1.046 healthy individuals for genotype and microbiome, these authors concluded that host genetics have a minor role in determining microbiome composition, contrasting with household sharing. They further stated that over 20% of the inter-person microbiome variability is associated with factors related to diet, drugs, and anthropometric measurements. These observations coincide with those from Lane et al. (2019), who, after studying the association between household composition and the infant faecal microbiome, concluded that the social environment of infants may influence the bacterial composition of the gastrointestinal infant microbiome (GIM). According to these authors, although the foundations of the GIM are likely evolutionarily conserved, dietary and environmental factors continuously modify it throughout an individual's lifespan.

The predominant role of diet claimed by these authors in defining gut microbiome composition would have an interesting potential cultural counterpart regarding the probability of a shared physiological background impact on social behavioural phenotypes, an issue that acquires important public consequences.

As commented by Amato (2016), based on data from previous authors, the host immune system is constantly monitoring the gut microbiota by dendritic cells and intestinal epithelial cells through extra-cellular matrix receptors and interacting with the gut-associated lymphoid tissue. Depending on these interactions, hosts can then affect gut microbial community composition.

In modern times, as has been stressed by Dominguez-Bello et al. (2019), selective pressures are shaping microbiome characteristics within high-income countries. These may include prenatal and postnatal antibiotics exposure, dietary antimicrobials, toothpaste, soaps, and perhaps even consumption of chlorinated water. These cultural factors operating on the microbiome can affect health conditions, susceptibility to noxa, and behavioural profiles, considering previous reports on brain-microbiome interactions. In this regard, as stated by Tito et al. (2012),

> Our results suggest that the most dramatic change to the gut microbiome in the human ancestral line has been the modern transformation of the human condition in cosmopolitan populations.

These results must call the attention of state and private educational programmers to include screening home feeding conditions, especially in children raised in impoverished settings (Colombo 2007; Lipina and Colombo 2009) and from families less aware of the impact of a complete diet on health and cognitive development. More yet, considering that several studies suggest that select probiotic treatments can modulate learning and memory behaviour in experimental animals (Vuong et al. 2017). Public health policies should include practical applications regarding these issues.

*

As mentioned in Allison et al. (2021) and as commented previously, diet is arguably the most influential lifestyle determinant of intestinal microbiota composition. Specific dietary habits, such as long-term adherence to a vegan/vegetarian diet or the consumption of Western vs. Mediterranean diets, are associated with distinct microbial profiles. In turn, the intestinal microbiota determines the production of bioactive metabolites, which regulate DNA methylation patterns in the adult brain and affect cognitive performance. In this domain, according to the authors, microbial regulation of cognitive behaviour would be mediated not only by pathways originating within the enteric nervous system but also through direct interactions between the central nervous system and microbe-derived metabolites. Metabolites of exclusively microbial origin and those produced due to host-microbiota combinatorial metabolism can cross the blood–brain barrier in mice, according to Swann et al. (2020).

As quoted in McDade et al. (2019), lower socioeconomic status (SES) is associated with physiological processes that contribute to disease development. For many non-human primates, social rank has also significantly impacted physiology and health. Within this domain, epigenetic processes may serve as important mechanisms of plasticity through which socioeconomic environments may generate child behavioural dysregulation and leave a molecular imprint that has lasting effects on the phenotype, as mentioned by Waterland and Michels (2007), Champagne (2010), Flannery et al. (2020) and others. DNA methylation is an important mechanism through which SES becomes biologically embedded across a large proportion of the genome, involving SES-sensitive periods and DNA methylation (Borghol et al. 2012; Lam et al. 2012; Needham et al. 2015).

Significant variations in cultural and regional feeding profiles anticipate differences in the gut microbiome, which would provide variable interactive behavioural profiles and drives, as implied in Ley et al. (2006) proposal. Studies on isolated human tribes such as the Amerindian Yanomami in Venezuela, with no documented previous contact with Western people, showed that they harbour a microbiome with the highest diversity of bacteria and genetic functions ever reported in a human group (Clemente et al., 2015). Authors suggested that the isolation and lack of transculturation of the Yanomami can also account for their higher bacterial diversity and prevalence. Kolodziejczyk et al. (2019) reported on the community of Hadza hunter-gatherers in Tanzania, which undergo cyclic changes in human gut microbiota due to seasonal variation in diet. However, consuming mainly raw or wild foods results in more diverse gut microbiota than the Western population. In turn, urbanisation would be associated with changes in composition, loss of diversity and loss of microbiome species, though, according to these authors, in comparison to simpler and more homogenous diets in rural areas, urban environments offer a large variety of foods, which leads to greater inter-individual variability of gut microbiomes.

The effects of a particular diet on individuals in the population differ from person to person and are influenced by a combination of host and microbiome features. The latter is determined mainly by the sociocultural environment rather than genetic background due to sociocultural profiles conditioning quality and variety in food consumption. Another variable associated with socioeconomic and cultural status is the consumption of saturated fatty acids, causing reduced microbiota richness and diversity in both adults and infants. It seems opportune to stress that these biases in the predominant consumption of carbohydrates and fatty acids significantly impact general nutrition and impair physical and neuro-behavioural development indexes at early developmental ages in impoverished populations.

*

As quoted by Alcock (2014), the modern feeding diet differentiates from our evolutionary ancestors in salt, simple carbohydrates, and saturated fat compared to the typical Western diet. This has been cited as the source of "diseases of civilisation," including obesity, cancer, and cardiovascular disease. Food preferences are thought to arise from interactions among genes, environment, and culture. At variance with modern food consumption, the human ancestral diet is thought to have contained foods far lower in salt, simple carbohydrates, and saturated fat than the Western diet. This discordance, or environmental mismatch, has been cited as the source of "diseases of civilisation". In this regard, Yatsunenko et al. (2012) characterised bacterial species present in faecal samples obtained from 531 individuals representing healthy Amerindians from the Amazonas of Venezuela, residents of rural Malawian communities, and inhabitants of USA metropolitan areas, as well as the gene content of 110 of their microbiomes. Pronounced differences were observed in bacterial species assemblages and functional gene repertoires between individuals residing in the USA compared to the other two countries. These distinctive features are evident in early infancy as well as adulthood. According to the authors, despite the large influence of cultural factors on which microbes are present in children and adults in each population, the basis for the similarity among family members was consistent across the three populations studied. Similar studies performed on children from indigenous and urban communities (i.e., westernised and non-westernised) in Mexico showed higher total diversity in the former population, as reported by Sanchez-Quinto et al. (2020). These authors further considered that,

> Since the industrial revolution, there have been numerous diet and life practice changes in Westernised communities, which have led to decreasing microbial diversity. This decrease affects the microbial enzymatic capacity for degrading nutrients and many forms of complex polysaccharides in human diets. Additionally, the depletion of microbial diversity can be associated with a broad range of inflammatory diseases, such as allergies, asthma, inflammatory bowel disease, obesity, and associated non-communicable diseases.

These authors also commented that the Mc'phaa people from a region known as "Montaña Alta" (*high mountain*) in the Mexican state of Guerrero are one of the groups whose lifestyle differs most strongly from the "Westernised lifestyle" typical of more urbanised areas.

Comparison of populations from two Oceania regions provided further knowledge of the selective pressures affecting microbial eco-systemic composition throughout human evolution and the potential consequences of pathophysiological states correlated with Westernisation lifestyles (Horwood et al. 2019). The main differences between these two communities were bacteria associated with different diets (high animal protein and refined sugars vs. high fibre food, respectively).

Dominguez-Bello et al. (2019) summarised that the microbiota ecosystems develop restricted to their epithelial niches by the host immune system, concomitantly with the host chronological development. Alcock et al. (2014) proposed that gut microbes may manipulate host eating behaviour to promote their fitness, even at the expense of host fitness. Norris et al. (2012) considered this issue on an evolutive domain, given that millions of years of coevolution of bacteria and their hosts have presumably selected those bacteria that best manipulate their hosts, generating a functional interaction between them. These authors proposed a hypothesis according to which there is a mutual reinforcement between the behaviour of the human host and the bacterial population within that host. In other words, a positive feedback loop exists between the host's preferences for a particular dietary regimen, the composition of the gut microbiota that depends on this regimen, and the host's preferences as influenced by the gut microbiota.

Involved in the mentioned interactions is the enteric nervous system with a complex web of microcircuits, neurotransmitters, and neuromodulators, which provided Gershon (1999) reasons to consider it a second brain, able to perform a complex, integrated set of functional processes. As discussed in other sections of this chapter, these processes appear to be associated with brain microstructure and cognitive function (Fernandez-Real et al. 2015).

Additionally, Shapira (2016) and Dominguez-Bello et al. (2019) stated that,

> The gut microbiome (the collection of microbial genomes) offers increased genetic variation compared with the host genome, faster evolution, and the ability to exchange microbes (and their genes and associated functions) with the environment, features that could contribute to host adaptation.
>
> *(Shapira 2016)*

> The microbiota has been transferred throughout generations of humans, with the matrilineal line transferring the primordial birth microbiota… The vertical human transmission has led to conservation of a phylogenetic signal in human microbiota communities.
>
> *(Dominguez-Bello et al. 2019)*

Though the molecular basis of microbe-host interactions and the roles of individual bacterial species remain to be fully elucidated, the microbial diversity (microbiome) of the human gut would be the result of coevolution between microbial *communities,* a consortium of gut microbe and their hosts (Ley et al. 2006) with which have evolved an intimate symbiotic relationship. As stated by these authors, defining the gut microbiota and the microbiome in people who live in various geographic regions under various levels of economic development should provide an opportunity to monitor human "microevolution" during this period of profound social, economic, and ecological change. In this domain, Kolodziejczyk et al. (2019) reported observations on the effects of diet on shaping the gut microbiota of Hadza hunter-gatherers in Tanzania and the Hutterites, an isolated, communal-living population in North America, based on seasonal variations. Finally, another critical factor driving dietary changes and subsequent microbiome alterations would be urbanisation, with a comparative loss of microbiome diversity.

References

Aagaard, Kjersti M. "Author Response to Comment on 'The Placenta Harbors a Unique Microbiome'." *Science Translational Medicine,* vol. 6, no. 254, 2014, doi:10.1126/scitranslmed.3010007.

Aatsinki, Anna-Katariina, *et al.* "Maternal Prenatal Psychological Distress and Hair Cortisol Levels Associate with Infant Fecal Microbiota Composition at 2.5 Months of Age." *Psychoneuroendocrinology,* vol. 119, 2020, p. 104754, doi:10.1016/j.psyneuen.2020.104754.

Aiello, Leslie C., and Peter Wheeler. "The Expensive-Tissue Hypothesis: The Brain and the Digestive System in Human and Primate Evolution." *Current Anthropology*, vol. 36, no. 2, 1995, pp. 199–221, doi:10.1086/204350.

Alcock, Joe, *et al.* "Is Eating Behavior Manipulated by the Gastrointestinal Microbiota? Evolutionary Pressures and Potential Mechanisms." *BioEssays*, vol. 36, no. 10, 2014, pp. 940–949, doi:10.1002/bies.201400071.

Allison, Joseph, *et al.* "Targeting DNA Methylation in the Adult Brain Through Diet." *Nutrients*, vol. 13, no. 11, 2021, p. 3979, doi:10.3390/nu13113979.

Amato, Katherine R. "Incorporating the Gut Microbiota into Models of Human and Non-Human Primate Ecology and Evolution." *American Journal of Physical Anthropology*, vol. 159, no. S61, 2016, pp. 196–215, doi:10.1002/ajpa.22908.

Arrieta, Marie-Claire, *et al.* "Early Infancy Microbial and Metabolic Alterations affect Risk of Childhood Asthma." *Science Translational Medicine*, vol. 7, no. 307, 2015, doi:10.1126/scitranslmed.aab2271.

Bäckhed, F. "Host-Bacterial Mutualism in the Human Intestine." *Science*, vol. 307, no. 5717, 2005, pp. 1915–1920, doi:10.1126/science.1104816.

Bäckhed, Fredrik, *et al.* "Dynamics and Stabilization of the Human Gut Microbiome during the First Year of Life." *Cell Host & Microbe*, vol. 17, no. 5, 2015, pp. 690–703, doi:10.1016/j.chom.2015.04.004.

Bailey, Michael T., *et al.* "Prenatal Stress alters Bacterial Colonization of the Gut in Infant Monkeys." *Journal of Pediatric Gastroenterology and Nutrition*, vol. 38, no. 4, 2004, pp. 414–421, doi:10.1097/00005176-200404000-00009.

Blekhman, Ran, *et al.* "Host Genetic Variation Impacts Microbiome Composition across Human Body Sites." *Genome Biology*, vol. 16, no. 1, 2015, doi:10.1186/s13059-015-0759-1.

Bocquet-Appel, Jean-Pierre. "Paleoanthropological Traces of a Neolithic Demographic Transition." *Current Anthropology*, vol. 43, no. 4, 2002, pp. 637–650, doi:10.1086/342429.

Bogaert, Debby, *et al.* "Mother-to-Infant Microbiota Transmission and Infant Microbiota Development across Multiple Body Sites." *Cell Host & Microbe*, vol. 31, no. 3, 2023, pp. 447–460, doi:10.1016/j.chom.2023.01.018.

Borghol, Nada, *et al.* "Associations with Early-Life Socio-Economic Position in Adult DNA Methylation." *International Journal of Epidemiology*, vol. 41, no. 1, 2012, pp. 62–74, doi:10.1093/ije/dyr147.

Champagne, Frances A. "Epigenetic Influence of Social Experiences Across the Lifespan." *Developmental Psychobiology*, vol. 52, no. 4, 2010, pp. 299–311, doi:10.1002/dev.20436.

Clapp, Megan, *et al.* "Gut Microbiota's Effect on Mental Health: The Gut-Brain Axis." *Clinics and Practice*, vol. 7, no. 4, 2017, p. 987, doi:10.4081/cp.2017.987.

Clarke, Gerard, *et al.* "Minireview: Gut Microbiota: The Neglected Endocrine Organ." *Molecular Endocrinology*, vol. 28, no. 8, 2014, pp. 1221–1238, www.ncbi.nlm.nih.gov/pmc/articles/PMC5414803/, doi:10.1210/me.2014-1108.

Clemente, Jose C., *et al.* "The Microbiome of Uncontacted Amerindians." *Science Advances*, vol. 1, no. 3, 2015, p. e1500183, advances.sciencemag.org/content/1/3/e1500183.full, doi:10.1126/sciadv.1500183.

Colombo, Jorge A. *Pobreza y Desarrollo Infantil. Una Contribucion Multidisciplinaria.* Buenos Aires: Ediciones Paidós, 2007, pp. 97–113.

Colombo, Jorge A. *Our Animal Condition and Social Construction.* New York: Nova Science Publishers Inc., 2019, eBook ISBN: 978-971-53615-53583.

Colombo, Jorge A. *Dominance Behavior: An Evolutive and Comparative Perspective.* Cham: Springer International Publishing, 2022.

Contijoch, Eduardo J., *et al.* "Gut Microbiota Density Influences Host Physiology and is Shaped by Host and Microbial Factors." *ELife*, vol. 8, 2019, doi:10.7554/elife.40553.

Costello, E. K., *et al.* "The Application of Ecological Theory Toward an Understanding of the Human Microbiome." *Science*, vol. 336, no. 6086, 2012, pp. 1255–1262, doi:10.1126/science.1224203.

Coyte, Katharine Z., *et al.* "The Ecology of the Microbiome: Networks, Competition, and Stability." *Science*, vol. 350, no. 6261, 2015, pp. 663–666, science.sciencemag.org/content/350/6261/663.full, doi:10.1126/science.aad2602.

Coyte, Katharine Z., *et al.* "Ecological Rules for the Assembly of Microbiome Communities." *PLOS Biology*, vol. 19, no. 2, 2021, p. e3001116, doi:10.1371/journal.pbio.3001116.

Cussotto, Sofia, *et al.* "The Neuroendocrinology of the Microbiota-Gut-Brain Axis: A Behavioural Perspective." *Frontiers in Neuroendocrinology*, vol. 51, 2018, pp. 80–101, doi:10.1016/j.yfrne.2018.04.002.

Dalby, Matthew J., and Lindsay J. Hall. "Recent Advances in Understanding the Neonatal Microbiome." *F1000Research*, vol. 9, 2020, p. 422, doi:10.12688/f1000research.22355.1.

Dinan, Timothy G., *et al.* "Collective Unconscious: How Gut Microbes Shape Human Behavior." *Journal of Psychiatric Research*, vol. 63, 2015, pp. 1–9, doi:10.1016/j.jpsychires.2015.02.021.

Dominguez-Bello, Maria Gloria, *et al.* "Delivery Mode Shapes the Acquisition and Structure of the Initial Microbiota across Multiple Body Habitats in Newborns." *Proceedings of the National Academy of Sciences of the United States of America*, vol. 107, no. 26, 2010, pp. 11971–11975, www.ncbi.nlm.nih.gov/pubmed/20566857, doi:10.1073/pnas.1002601107.

Dominguez-Bello, Maria Gloria, *et al.* "Role of the Microbiome in Human Development." *Gut*, vol. 68, no. 6, 2019, pp. 1108–1114, gut.bmj.com/content/68/6/1108, doi:10.1136/gutjnl-2018-317503.

Douglas, Angela E. "Symbiosis as a General Principle in Eukaryotic Evolution." *Cold Spring Harbor Perspectives in Biology*, vol. 6, no. 2, 2014, pp. a016113–a016113, cshperspectives.cshlp.org/content/6/2/a016113.full, doi:10.1101/cshperspect.a016113.

Dunbar, Robin I. M., and Susanne Shultz. "Evolution in the Social Brain." *Science*, vol. 317, no. 5843, 2007, pp. 1344–1347, doi:10.1126/science.1145463.

Dunbar, Robin I. M. "The Social Brain Hypothesis and Its Implications for Social Evolution." *Annals of Human Biology*, vol. 36, no. 5, 2009, pp. 562–572, doi:10.1080/03014460902960289.

Evans, James M., *et al.* "The Gut Microbiome: The Role of a Virtual Organ in the Endocrinology of the Host." *Journal of Endocrinology*, vol. 218, no. 3, 2013, pp. R37–R47, doi:10.1530/joe-13-0131.

Fernandez-Real, José-Manuel, *et al.* "Gut Microbiota Interacts with Brain Microstructure and Function." *The Journal of Clinical Endocrinology & Metabolism*, vol. 100, no. 12, 2015, pp. 4505–4513, doi:10.1210/jc.2015-3076.

Ferretti, Pamela, *et al.* "Mother-To-Infant Microbial Transmission from Different Body Sites Shapes the Developing Infant Gut Microbiome." *Cell Host & Microbe*, vol. 24, no. 1, 2018, pp. 133–145.e5, www.ncbi.nlm.nih.gov/pmc/articles/PMC6716579/, doi:10.1016/j.chom.2018.06.005.

Flannery, Jessica E., *et al.* "Gut Feelings Begin in Childhood: The Gut Metagenome Correlates with Early Environment, Caregiving, and Behavior." *MBio*, vol. 11, no. 1, 2020, doi:10.1128/mbio.02780-19.

Foster, Kevin R., *et al.* "The Evolution of the Host Microbiome as an Ecosystem on a Leash." *Nature*, vol. 548, no. 7665, 2017, pp. 43–51, doi:10.1038/nature23292.

Funkhouser, Lisa J., and Seth R. Bordenstein. "Mom Knows Best: The Universality of Maternal Microbial Transmission." *PLoS Biology*, vol. 11, no. 8, 2013, www.ncbi.nlm.nih.gov/pmc/articles/PMC3747981/, doi:10.1371/journal.pbio.1001631.

Garud, Nandita R., *et al.* "Evolutionary Dynamics of Bacteria in the Gut Microbiome Within and Across Hosts." *PLoS Biology*, vol. 17, no. 1, 2019, p. e3000102, doi:10.1371/journal.pbio.3000102.

Gaufin, Thaidra, *et al.* "The Importance of the Microbiome in Pediatrics and Pediatric Infectious Diseases." *Current Opinion in Pediatrics*, vol. 30, no. 1, 2018, pp. 117–124, doi:10.1097/mop.0000000000000576.

Gensollen, Thomas, *et al.* "How Colonization by Microbiota in Early Life Shapes the Immune System." *Science*, vol. 352, no. 6285, 2016, pp. 539–544, www.ncbi.nlm.nih.gov/pmc/arti cles/PMC5050524/, doi:10.1126/science.aad9378.

Gershon, Michael D. "The Enteric Nervous System: A Second Brain." *Hospital Practice*, vol. 34, no. 7, 1999, pp. 31–52, doi:10.3810/hp.1999.07.153.

Gilbert, Scott F. "A Holobiont Birth Narrative: The Epigenetic Transmission of the Human Microbiome." *Frontiers in Genetics*, vol. 5, 2014, doi:10.3389/fgene.2014.00282.

Goodrich, Julia K., *et al.* "Human Genetics Shape the Gut Microbiome." *Cell*, vol. 159, no. 4, 2014, pp. 789–799, doi:10.1016/j.cell.2014.09.053.

Goodrich, Julia K., *et al.* "Genetic Determinants of the Gut Microbiome in UK Twins." *Cell Host & Microbe*, vol. 19, no. 5, 2016, pp. 731–743, doi:10.1016/j.chom.2016.04.017, PMID: 27173935, PMCID: PMC4915943.

Gordon, Helmut A., and Laszlo Pesti. "The Gnotobiotic Animal as a Tool in the Study of Host Microbial Relationships." *Bacteriological Reviews*, vol. 35, no. 4, 1971, pp. 390–429, doi:10.1128/br.35.4.390-429.1971.

Goulet, Olivier. "Potential Role of the Intestinal Microbiota in Programming Health and Disease: Figure 1." *Nutrition Reviews*, vol. 73, no. suppl. 1, 2015, pp. 32–40. doi:10.1093/nutrit/nuv039.

Hechler, C., *et al.* "Association between Psychosocial Stress and Fecal Microbiota in Pregnant Women." *Scientific Reports*, vol. 9, no. 1, 2019, doi:10.1038/s41598-019-40434-8.

Heijtz, R. D., *et al.* "Normal Gut Microbiota Modulates Brain Development and Behavior." *Proceedings of the National Academy of Sciences*, vol. 108, no. 7, 2011, pp. 3047–3052, doi:10.1073/pnas.1010529108.

Henry, Lucas P., *et al.* "The Microbiome Extends Host Evolutionary Potential." *Nature Communications*, vol. 12, no. 1, 2021, p. 5141, www.nature.com/articles/s41467-021-25315-x, doi:10.1038/s41467-021-25315-x.

Hitze, B., *et al.* "How the Selfish Brain Organizes its Supply and Demand." *Frontiers in Neuorenergetics*, vol. 2, no. 7, 2010, pp. 1–13, doi:10.3389/fnene.2010.00007.

Horwood, Paul F.*et al.* "Health Challenges of the Pacific Region: Insights from History, Geography, Social Determinants, Genetics, and the Microbiome." *Frontiers in Immunology*, vol. 10, 2019, doi:10.3389/fimmu.2019.02184.

House, Patrick K., *et al.* "Predator Cat Odors Activate Sexual Arousal Pathways in Brains of Toxoplasma Gondii Infected Rats." *PLoS ONE*, vol. 6, no. 8, 2011, p. e23277, doi:10.1371/journal.pone.0023277.

Hughes, David T., and Vanessa Sperandio. "Inter-Kingdom Signalling: Communication between Bacteria and Their Hosts." *Nature Reviews Microbiology*, vol. 6, no. 2, 2008, pp. 111–120, www.ncbi.nlm.nih.gov/pmc/articles/PMC2667375/, doi:10.1038/nrmicro1836.

Jansma, Jack, and Sahar El Aidy. "Understanding the Host-Microbe Interactions Using Metabolic Modeling." *Microbiome*, vol. 9, no. 16, 2021, doi:10.1186/s40168-020-00955-1.

Johnson, Katerina V. A. "Gut Microbiome Composition and Diversity Are Related to Human Personality Traits." *Human Microbiome Journal*, vol. 15, 2019, p. 100069, www.sciencedir ect.com/science/article/pii/S2452231719300181, doi:10.1016/j.humic.2019.100069.

Kim, Joon Yong, *et al.* "The Human Gut Archaeome: Identification of Diverse Haloarchaea in Korean Subjects." *Microbiome*, vol. 8, no. 1, 2020, doi:10.1186/s40168-020-00894-x.

Koenig, Jeremy E., *et al.* "Succession of Microbial Consortia in the Developing Infant Gut Microbiome." *Proceedings of the National Academy of Sciences of the United States of America*, vol. 108, Suppl. 1, 2011, pp. 4578–4585, www.ncbi.nlm.nih.gov/pubmed/20668239, doi:10.1073/pnas.1000081107.

Kolodziejczyk, Aleksandra A., *et al.* "Diet–Microbiota Interactions and Personalized Nutrition." *Nature Reviews Microbiology*, vol. 17, no. 12, 2019, pp. 742–753, www.nature.com/articles/s41579-019-0256-8, doi:10.1038/s41579-019-0256-8.

Kurokawa, Ken, *et al.* "Comparative Metagenomics Revealed Commonly Enriched Gene Sets in Human Gut Microbiomes." *DNA Research*, vol. 14, no. 4, 2007, pp. 169–181, doi:10.1093/dnares/dsm018.

Kurokawa, Shunya, *et al.* "The Effect of Fecal Microbiota Transplantation on Psychiatric Symptoms among Patients with Irritable Bowel Syndrome, Functional Diarrhea and Functional Constipation: An Open-Label Observational Study." *Journal of Affective Disorders*, vol. 235, 2018, pp. 506–512, pubmed.ncbi.nlm.nih.gov/29684865/, doi:10.1016/j.jad.2018.04.038.

Lam, Lucia L., *et al.* "Factors Underlying Variable DNA Methylation in a Human Community Cohort." *Proceedings of the National Academy of Sciences*, vol. 109, no. suppl. 2, 2012, pp. 17253–17260, doi:10.1073/pnas.1121249109.

Lane, Avery A., *et al.* "Household Composition and the Infant Fecal Microbiome: The INSPIRE Study." *American Journal of Physical Anthropology*, vol. 169, no. 3, 2019, pp. 526–539, doi:10.1002/ajpa.23843.

Leonard, William R., *et al.* "Metabolic Correlates of Hominid Brain Evolution." *Comparative Biochemistry and Physiology Part A: Molecular & Integrative Physiology*, vol. 136, no. 1, 2003, pp. 5–15, doi:10.1016/s1095-6433(03)00132-6.

Leonard, William R., *et al.* "Effects of Brain Evolution on Human Nutrition and Metabolism." *Annual Review of Nutrition*, vol. 27, no. 1, 2007, pp. 311–327, doi:10.1146/annurev.nutr.27.061406.093659.

Ley, Ruth E., *et al.* "Ecological and Evolutionary Forces Shaping Microbial Diversity in the Human Intestine." *Cell*, vol. 124, no. 4, 2006, pp. 837–848, doi:10.1016/j.cell.2006.02.017.

Ley, Ruth E., *et al.* "Worlds Within Worlds: Evolution of the Vertebrate Gut Microbiota." *Nature Reviews Microbiology*, vol. 6, no. 10, 2008a, pp. 776–788, www.ncbi.nlm.nih.gov/pubmed/18794915, doi:10.1038/nrmicro1978.

Ley, Ruth E., *et al.* "Evolution of Mammals and Their Gut Microbes." *Science*, vol. 320, no. 5883, 2008b, pp. 1647–1651, www.ncbi.nlm.nih.gov/pmc/articles/PMC2649005/, doi:10.1126/science.1155725.

Li, M., *et al.* "Symbiotic Gut Microbes Modulate Human Metabolic Phenotypes." *Proceedings of the National Academy of Sciences*, vol. 105, no. 6, 2008, pp. 2117–2122, doi:10.1073/pnas.0712038105.

Lipina, Sebastián J., and Jorge A. Colombo. *Poverty and Brain Development during Childhood: An Approach from Cognitive Psychology and Neuroscience.* Washington, D.C.: American Psychological Association, 2009.

Lombardo, Michael P. "Access to Mutualistic Endosymbiotic Microbes: An Underappreciated Benefit of Group Living." *Behavioral Ecology and Sociobiology*, vol. 62, no. 4, 2007, pp. 479–497, doi:10.1007/s00265-007-0428-9.

López-Goñi, Ignacio. *Microbiota: Los Microbios de Tu Organismo.* Cordoba: Editorial Almuzara, 2018.

Lyte, Mark. "The Microbial Organ in the Gut as a Driver of Homeostasis and Disease." *Medical Hypotheses*, vol. 74, no. 4, 2010, pp. 634–638, doi:10.1016/j.mehy.2009.10.025.

Mackie, Roderick I., *et al.* "Developmental Microbial Ecology of the Neonatal Gastrointestinal Tract." *The American Journal of Clinical Nutrition*, vol. 69, no. 5, 1999, pp. 1035s–1045s, doi:10.1093/ajcn/69.5.1035s.

Macnab, Vicki, and Iain Barber. "Some (Worms) like It Hot: Fish Parasites Grow Faster in Warmer Water, and Alter Host Thermal Preferences." *Global Change Biology*, vol. 18, no. 5, 2011, pp. 1540–1548, doi:10.1111/j.1365-2486.2011.02595.x.

McBurney, Michael I., *et al.* "Establishing What Constitutes a Healthy Human Gut Microbiome: State of the Science, Regulatory Considerations, and Future Directions." *The Journal of Nutrition*, vol. 149, no. 11, 2019, pp. 1882–1895, doi:10.1093/jn/nxz154.

McDade, Thomas W., *et al.* "Genome-Wide Analysis of DNA Methylation in Relation to Socioeconomic Status during Development and Early Adulthood." *American Journal of Physical Anthropology*, vol. 169, no. 1, 2019, pp. 3–11, doi:10.1002/ajpa.23800.

McFall-Ngai, Margaret, *et al.* "Animals in a Bacterial World, a New Imperative for the Life Sciences." *Proceedings of the National Academy of Sciences*, vol. 110, no. 9, 2013, pp. 3229–3236, doi:10.1073/pnas.1218525110.

Mira, Alex, *et al.* "The Neolithic Revolution of Bacterial Genomes." *Trends in Microbiology*, vol. 14, no. 5, 2006, pp. 200–206, doi:10.1016/j.tim.2006.03.001.

Moeller, Andrew H., *et al.* "Rapid Changes in the Gut Microbiome During Human Evolution." *Proceedings of the National Academy of Sciences*, vol. 111, no. 46, 2014, pp. 16431–16435, doi:10.1073/pnas.1419136111.

Moeller, Andrew H., and Jon G. Sanders. "Roles of the Gut Microbiota in the Adaptive Evolution of Mammalian Species." *Philosophical Transactions of the Royal Society B: Biological Sciences*, vol. 375, no. 1808, 2020, p. 2019.0597, doi:10.1098/rstb.2019.0597.

Moore, Rebecca E., and Steven D. Townsend. "Temporal Development of the Infant Gut Microbiome." *Open Biology*, vol. 9, no. 9, 2019, p. 190128, doi:10.1098/rsob.190128.

Moossavi, Shirin, and Meghan B. Azad. "Origins of Human Milk Microbiota: New Evidence and Arising Questions." *Gut Microbes*, vol. 12, no. 1, 2019, pp. 1–10, doi:10.1080/19490976.2019.1667722.

Moran, Nancy A., *et al.* "Evolutionary and Ecological Consequences of Gut Microbial Communities." *Annual Review of Ecology, Evolution and Systematics*, vol. 50, no. 1, 2019, pp. 451–475, doi:10.1146/annurev-ecolsys-110617-062453.

Needham, Belinda L., *et al.* "Life Course Socioeconomic Status and DNA Methylation in Genes Related to Stress Reactivity and Inflammation: The Multi-Ethnic Study of Atherosclerosis." *Epigenetics*, vol. 10, no. 10, 2015, pp. 958–969, doi:10.1080/15592294.2015.1085139.

Nishida, Alex H., and Howard Ochman. "Rates of Gut Microbiome Divergence in Mammals." *Molecular Ecology*, vol. 27, no. 8, 2018, pp. 1884–1897, doi:10.1111/mec.14473.

Nishida, Alex H., and Howard Ochman. "A Great-Ape View of the Gut Microbiome." *Nature Reviews Genetics*, vol. 20, no. 4, 2019, pp. 195–206, www.nature.com/articles/s41576-018-0085-z, doi:10.1038/s41576-018-0085-z.

Norris, Victor, *et al.* "Hypothesis: Bacteria Control Host Appetites." *Journal of Bacteriology*, vol. 195, no. 3, 2012, pp. 411–416, doi:10.1128/jb.01384-12.

Ochman, Howard, *et al.* "Evolutionary Relationships of Wild Hominids Recapitulated by Gut Microbial Communities." *PLoS Biology*, vol. 8, no. 11, 2010, p. e1000546, doi:10.1371/journal.pbio.1000546.

Org, Elin, *et al.* "Genetic and Environmental Control of Host-Gut Microbiota Interactions." *Genome Research*, vol, 25, no. 10, 2015, pp. 1558–1569, doi:10.1101/gr.194118.115.

Parashar, Arun, and Malairaman Udayabanu. "Gut Microbiota Regulates Key Modulators of Social Behavior." *European Neuropsychopharmacology*, vol. 26, no. 1, 2016, pp. 78–91. doi:10.1016/j.euroneuro.2015.11.002.

Penders, John, *et al.* "Factors Influencing the Composition of the Intestinal Microbiota in Early Infancy." *Pediatrics*, vol. 118, no. 2, 2006, pp. 511–521, www.ncbi.nlm.nih.gov/pubmed/16882802, doi:10.1542/peds.2005-2824.

Perez-Muñoz, Maria Elisa, *et al.* "A Critical Assessment of the 'Sterile Womb' and in Utero Colonization' Hypotheses: Implications for Research on the Pioneer Infant Microbiome." *Microbiome*, vol. 5, no. 1, 2017, doi:10.1186/s40168-017-0268-4, PMID: 28454555; PMCID: PMC5410102.

Peters, A., *et al.* "The Selfish Brain: Competition for Energy Resources." *Neuroscience & Biobehavioral Reviews*, vol. 28, no. 2, 2004, pp. 143–180, doi:10.1016/j.neubiorev.2004.03.002.

Previc, Fred H. "Dopamine and the Origins of Human Intelligence." *Brain and Cognition*, vol. 41, no. 3, 1999, pp. 299–350, www.sciencedirect.com/science/article/abs/pii/S02782626 99911296?via%3Dihub, doi:10.1006/brcg.1999.1129.

Previc, Fred H. "The Role of the Extrapersonal Brain Systems in Religious Activity." *Consciousness and Cognition*, vol. 15, no. 3, 2006, pp. 500–539, doi:10.1016/j.concog.2005.09.009.

Previc, Fred H. *Dopaminergic Mind in Human Evolution and History.* Cambridge, UK: Cambridge University Press, 2009.

Pronovost, Geoffrey N., and Elaine Y. Hsiao. "Perinatal Interactions between the Microbiome, Immunity, and Neurodevelopment." *Immunity*, vol. 50, no. 1, 2019, pp. 18–36, doi:10.1016/j.immuni.2018.11.016.

Qian, Jimmy J., and Erol Akçay. "The Balance of Interaction Types Determines the Assembly and Stability of Ecological Communities." *Nature Ecology & Evolution*, vol. 4, no. 3, 2020, pp. 356–365, doi:10.1038/s41559-020-1121-x.

Qin, Junjie, *et al.* "A Human Gut Microbial Gene Catalogue Established by Metagenomic Sequencing." *Nature*, vol. 464, no. 7285, 2010, pp. 59–65, doi:10.1038/nature08821.

Rodriguez, Juan Miguel, *et al.* "The Composition of the Gut Microbiota throughout Life, with an Emphasis on Early Life." *Microbial Ecology in Health & Disease*, vol. 26, 2015, doi:10.3402/mehd.v26.26050.

Rothschild, Daphna, *et al.* "Environment Dominates over Host Genetics in Shaping Human Gut Microbiota." *Nature*, vol. 555, no. 7695, 2018, pp. 210–215. doi:10.1038/nature25973.

Sampson, Timothy R., and Sarkis K. Mazmanian. "Control of Brain Development, Function, and Behavior by the Microbiome." *Cell Host & Microbe*, vol. 17, no. 5, 2015, pp. 565–576, doi:10.1016/j.chom.2015.04.011.

Sánchez-Quinto, Andrés, *et al.* "Gut Microbiome in Children from Indigenous and Urban Communities in México: Different Subsistence Models, Different Microbiomes." *Microorganisms*, vol. 8, no. 10, 2020, p. 1592, www.mdpi.com/2076-2607/8/10/1592, doi:10.3390/microorganisms8101592.

Sarkar, Amar, *et al.* "The Role of the Microbiome in the Neurobiology of Social Behaviour." *Biological Reviews*, vol. 95, no. 5, 2020, pp. 1131–1166, doi:10.1111/brv.12603.

Savage, D. C. "Gastrointestinal Microflora in Mammalian Nutrition." *Annual Review of Nutrition*, vol. 6, no. 1, 1986, pp. 155–178, doi:10.1146/annurev.nu.06.070186.001103.

Shapira, Michael. "Gut Microbiotas and Host Evolution: Scaling up Symbiosis." *Trends in Ecology & Evolution*, vol. 31, no. 7, 2016, pp. 539–549, doi:10.1016/j.tree.2016.03.006.

Sharon, Gil, *et al.* "The Central Nervous System and the Gut Microbiome." *Cell*, vol. 167, no. 4, 2016, pp. 915–932, doi:10.1016/j.cell.2016.10.027.

Sharon, Gil, *et al.* "Human Gut Microbiota from Autism Spectrum Disorder Promote Behavioral Symptoms in Mice." *Cell*, vol. 177, no. 6, 2019, pp. 1600–1618.e17, www.sciencedirect.com/science/article/pii/S0092867419305021, doi:10.1016/j.cell.2019.05.004.

Steiger, Sandra, *et al.* "The Origin and Dynamic Evolution of Chemical Information Transfer." *Proceedings of the Royal Society B: Biological Sciences*, vol. 278, no. 1708, 2010, pp. 970–979, doi:10.1098/rspb.2010.2285.

Stewart, Christopher J., *et al.* "Temporal Development of the Gut Microbiome in Early Childhood from the TEDDY Study." *Nature*, vol. 562, no. 7728, 2018, pp. 583–588, www.nature.com/articles/s41586-018-0617-x, doi:10.1038/s41586-018-0617-x.

Stilling, Roman M., *et al.* "Friends with Social Benefits: Host-Microbe Interactions as a Driver of Brain Evolution and Development?" *Frontiers in Cellular and Infection Microbiology*, vol. 4, no. 147, 2014, doi:10.3389/fcimb.2014.00147.

Stilling, Roman M., *et al.* "The Brain's Geppetto—Microbes as Puppeteers of Neural Function and Behaviour?" *Journal of NeuroVirology*, vol. 22, no. 1, 2015, pp. 14–21, doi:10.1007/s13365-015-0355-x.

Stinson, Lisa F. "Establishment of the Early-Life Microbiome: A DOHaD Perspective." *Journal of Developmental Origins of Health and Disease*, vol. 11, no. 3, 2020, pp. 201–210, doi:10.1017/s2040174419000588.

Swann, Jonathan R., *et al.* "Developmental Signatures of Microbiota-Derived Metabolites in the Mouse Brain." *Metabolites*, vol. 10, no. 5, 2020, p. 172, doi:10.3390/metabo10050172.

Tito, Raul Y., *et al.* "Insights from Characterizing Extinct Human Gut Microbiomes." *PLoS ONE*, vol. 7, no. 12, 2012, p. e51146, doi:10.1371/journal.pone.0051146.

Trevelline, Brian K., and Kevin D. Kohl. "The Gut Microbiome Influences Host Diet Selection Behavior." *Proceedings of the National Academy of Sciences*, vol. 119, no. 17, 2022, doi:10.1073/pnas.2117537119.

Turnbaugh, Peter J., *et al.* "The Effect of Diet on the Human Gut Microbiome: A Metagenomic Analysis in Humanized Gnotobiotic Mice." *Science Translational Medicine*, vol. 1, no. 6, 2009, doi:10.1126/scitranslmed.3000322.

Valles-Colomer, Mireia, *et al.* "Variation and Transmission of the Human Gut Microbiota across Multiple Familial Generations." *Nature Microbiology*, vol. 7, no. 1, 2022, pp. 87–96, doi:10.1038/s41564-021-01021-8.

Vuong, Helen E., *et al.* "The Microbiome and Host Behavior." *Annual Review of Neuroscience*, vol. 40, no. 1, 2017, pp. 21–49, doi:10.1146/annurev-neuro-072116-031347.

Walter, Jens, and Ruth Ley. "The Human Gut Microbiome: Ecology and Recent Evolutionary Changes." *Annual Review of Microbiology*, vol. 65, no. 1, 2011, pp. 411–429, doi:10.1146/annurev-micro-090110-102830.

Waterland, Robert A., and Karin B. Michels. "Epigenetic Epidemiology of the Developmental Origins Hypothesis." *Annual Review of Nutrition*, vol. 27, no. 1, 2007, pp. 363–388, doi:10.1146/annurev.nutr.27.061406.093705.

Whiten, Andrew, and David Erdal. "The Human Socio-Cognitive Niche and Its Evolutionary Origins." *Philosophical Transactions of the Royal Society B: Biological Sciences*, vol. 367, no. 1599, 2012, pp. 2119–2129, doi:10.1098/rstb.2012.0114.

Yatsunenko, Tanya, *et al.* "Human Gut Microbiome Viewed across Age and Geography." *Nature*, vol. 486, no. 7402, 2012, pp. 222–227, doi:10.1038/nature11053.

Zijlmans, Maartje A. C., *et al.* "Maternal Prenatal Stress is Associated with the Infant Intestinal Microbiota." *Psychoneuroendocrinology*, vol. 53, 2015, pp. 233–245, doi:10.1016/j.psyneuen.2015.01.006.

7

BRAIN-GUT MICROBIOME INTERACTIONS

Animals cannot be considered individuals by anatomical or physiological criteria because a diversity of symbionts are both present and functional in completing metabolic pathways and serving other physiological functions. Similarly, these new studies have shown that animal development is incomplete without symbionts. Symbionts also constitute a second mode of genetic inheritance, providing selectable genetic variation for natural selection.

(Gilbert et al. 2012)

Through a network of microbe–microbe and host–microbe interactions, the gut microbiota is capable of substrate decomposition and metabolite biosynthesis pathways that no individual member can replicate in isolation. Moreover, considerable functional redundancy exists, whereby the same process can be performed by many different microbes, through many different pathways.

(Shoubridge et al. 2022)

Brain-Gut Microbiome Interactions

In a striking display of trans-kingdom symbiosis, gut bacteria cooperate with their animal hosts to regulate the development and function of the immune, metabolic and nervous systems through dynamic bidirectional communication along the "gut–brain axis". These processes may affect human health, as certain animal behaviours appear to correlate with the composition of gut bacteria, and disruptions in microbial communities have been implicated in several neurological disorders.

(Morais et al. 2021)

DOI: 10.4324/9781032698380-7

The co-evolution of animals and their associated microbial communities resulted in complex biological communications between the gut and the brain. Three significant ways the microbiota can influence the development and function of the nervous system include modulation of the immune response, impact on metabolism, including hormones, neuropeptides and neurotransmitters, and direct effects on neurons and neuronal signalling (Morais et al. 2021). These authors considered that gut microbiota communities at the intersection of the host and the environment act as a filter and biological rheostat for sensing, modifying, and tuning vast amounts of chemical signals from the environment that circulate throughout the body.

*

The blood-brain barrier develops during the early period of intrauterine life and is formed by capillary endothelial cells sealed by tight junctions, astrocytes, and pericytes, required for an optimal microenvironment for neuronal growth, cell specification, and normal brain function. According to Braniste et al. (2014), gut microbiota affects the permeability of the blood-brain barrier.

An interactive relationship has been explored relating to sociality, enhanced transmission of microbiota, and brain /mental evolution, as suggested by the following reports. Stilling et al. (2014), speculated that enhanced transmission of microbes through group living may have contributed to the gradual increase in cortical size and function. Luczynski et al. (2016) stressed the role of the microbiota in regulating the brain and behaviour, as their findings suggest that the gut microbiota is necessary for the normal dendritic cell morphology of the amygdala and hippocampus. Further, the authors suggested that microstructural changes could underline the altered stress responses and behavioural profiles observed in germ-free mice. They concluded that results in germ-free animals show altered brain morphology and suggested that the amygdala and hippocampus are brain regions whose structural integrity is contingent on the presence of gut microbiota, thus involving the expression of limbic system-mediated behaviours and physiology. Since the neural circuits involved in behavioural domains include prefrontal circuits, it is interesting Dunbar's (2018) comment that a positive relationship between neocortex size and social behaviour within primates is a factor critically contributing to the evolution of human intelligence. Novotny et al. (2019) considered that while the effect of dieting on neuro-behavioural functions has been studied for a long time, the microbiome's effect on human cognition is still partially unexplored. Recent correlation studies showed that the abundance of 14 interacting genera in the gut microbiome was positively linked to fluid intelligence score (Oluwagbemigun et al. 2022). This would support the concept that the gut microbiome may be involved in cognitive performance. If so, this concept would have a significant impact on nutritional public policies involving impoverished populations and its impact on early cognitive development.

As stated, the results mentioned are relevant to emotions and social behaviour. Davidson et al. (2018), remark that what makes the relationship between the microbiome and the brain especially significant from an evolutionary and ecological perspective is that research on humans and other animals, including insects, birds, and mammals, suggests that the microbiome differs among individuals due to environmental factors. This would include sociality and host genotype, adding that the microbiome can directly affect cognition, as demonstrated in laboratory mice and human infants. These authors further stated that experimental examination of the microbiome through diet, infection, stress, and exercise manipulation suggests direct effects on cognition, including learning and memory. In this regard, Villa and Sanchez-Perez (2021) considered that gut microorganisms can either produce or degrade chemical compounds involved in signalling pathways associated with happiness, depression, suicidal behaviours, or even aggressiveness. According to Meyer et al. (2022), these effects would extend to cognitive function in midlife, suggesting that the gut microbiota may be associated with cognitive ageing, though the authors consider that results must be replicated in larger samples.

In addition to studies by Tung et al. (2015) showing that social networks predict microbial composition in wild primates, Davidson et al. (2018) predicted that positions in social networks may also be affected by the microbiome, which affects the social transmission of microbes. These authors discussed the role of the microbiome as a driver of cognitive evolution as it has been theoretically explored within the context of the *social brain hypothesis* proposed by Dunbar (1998,2003).

> The social brain (or Machiavellian Intelligence) hypothesis was proposed to explain primates' unusually large brains: It argues that the cognitive demands of living in complexly bonded social groups selected for increases in executive brain (principally neocortex).
>
> *(Dunbar 2003)*

According to Stilling et al. (2014) selection would favour social complexity to increase the transmission of beneficial microbes, causing host-symbiont co-evolution on RNA regulation in the brain.

*

> Children living in poverty generally perform poorly in school, with markedly lower standardised test scores and lower educational attainment. The longer children live in poverty, the greater their academic deficits. These patterns persist to adulthood, contributing to lifetime-reduced occupational attainment .... The influence of poverty on children's learning and achievement is mediated by structural brain development.
>
> *(Hair et al. 2015)*

Microbiome colonisation would play a fundamental role in brain development in the early postnatal weeks. This issue was discussed by Heijtz et al. (2011), reporting on microbial colonisation process signalling mechanisms and suggesting that during evolution, gut microbiota colonisation has become integrated into the programming of brain development, affecting motor control and anxiety-like behaviour. Effects on brain development, ageing, and neurodegeneration were also discussed by Dinan and Cryan (2017). According to these authors, infancy is a critical microbiota colonisation and neurodevelopment period. Additionally, the effects of the microbiome on brain development would not be specific to the hippocampus, based on data indicating that there are alterations in amygdaline function as well.

Considering the nodal role of prefrontal neural circuits in neurodevelopment and normal brain/mental function, Hoban et al. (2016) published their observations on prefrontal myelination development in germ-free rodents. According to the authors, targeting the microbiota during critical periods of synaptic reconstruction (early adolescence post-weaning) significantly affected myelination at the transcriptional level. Results would suggest that appropriate cortical myelination of the prefrontal cortex in rodents relies on the presence of a functional microbiota during critical windows of neurodevelopment. These conclusions obviously must await comparative studies with species undergoing neuro-behavioural developmental timing different from the rodents, as in primates. Dinan and Cryan (2016) quoted preclinical studies primarily based on rodents, showing that the microbiome is vital to normal neurodevelopment and behaviour. The brain fails to develop normally without the gut microbiome in germ-free animals. As quoted by these authors, fundamental neural processes such as myelination, adult neurogenesis, and microglia activation have also shown dependence on microbiota composition.

Dinan et al. (2015) considered that humans fundamentally depend on a myriad of essential neurochemicals produced by microbes. Since multicellular life emerged from unicellular life forms during evolution, the latter remained dominant on the planet or as symbiotic or parasitic relationships in multicellular organisms. Regarding its relationship with brain development, the authors propose that,

> ... the development of a complex gut microbiota in mammals has played an important role in enabling brain development, especially in terms of cognitive function and fundamental behaviour patterns, such as facilitating social interaction and effectively dealing with environmental stressors.

According to Smith and Wissel (2019), though research on microbe-host interactions has predominantly been conducted with nonhuman models, it may be time to expand the concept of "body" to include the microorganisms that live on and inside humans and explore its neuro-behavioural derivations.

Considering the potential impact on timing and degree of neuro-cognitive development, the above concepts project onto public policy domains regarding food access and feeding profiles, especially in impoverished communities and at early developmental ages, as will be further discussed.

Has social behaviour undergone positive selection under this host-microbiome co-evolution to enhance the probability of microbial transmission?

*

The development of financial and sociopolitical structures has led *Homo sapiens* to construct a segmented world community, affecting educational access and cognitive processes, health security involving feeding insecurity, distorting access to critical knowledge, and increasing ecological risk. Among other negative social impacts, these interactive conditions interfere with optimal cognitive development and identity construction, thus limiting individual competitiveness in adult life. Struggling to have access to basic survival needs affects the chances of an optimised individual and community development and places their survival at risk in terms of physical and cognitive domains. These conditions acquire a dramatic character in those communities exposed to poverty and indigence, often lasting several generations. These communities do not belong to socially integrated, preindustrial cultures but to the condition of outsiders of the modern world.

According to Penders et al. (2006), the most critical determinants of the gut microbiota composition in infants would be the mode of child delivery, type of infant feeding, gestational age, infant hospitalisation, and infant antibiotic treatment. It was underlined that infants born vaginally at home and breastfed exclusively seemed to have the most "beneficial" gut microbiota.

As considered by Münger et al. (2018), gestation and infancy appear to represent critical periods for the gut microbiome to influence the infant's brain development and may be critical in determining its future degree of sociability. Laue et al. (2022) stated that studies have highlighted the birth window to three years as a susceptible window when microbiome interventions may be the most effective and further considered that the foundation for lifelong health is laid during this early fundamental window. According to Roswall et al. (2021), several bacterial taxa that have been associated with human health are acquired late in childhood and have not reached their adult abundance at five years of age and would develop at an individual pace along the microbiota's developmental trajectory. Furthermore, the authors considered that maternal microbiome and metabolome play a role in foetal neurodevelopment and subsequent labilities due to the scope of processes modulated by the microbiome. Based on data from experimental animal models (rodents), the authors considered that the microbiome modulates basic neurodevelopmental processes, such as blood-brain barrier formation, neurogenesis, microglia maturation and the expression of neurotrophins and neurotransmitters, and their receptors.

Tooley (2020) reported that though comparatively reduced data are available from human trials, correlations between microbiota diversity and enhanced cognitive flexibility and executive function were observed. The author reported that,

> Limited studies were available to draw a detailed conclusion; however, available evidence suggests that gut microbiota is linked to cognitive performance and that manipulation of gut microbiota could be a promising avenue for enhancing cognition which warrants further research.

These results tend to coincide with those later obtained by Rothenberg et al. (2021) in children from rural China, though the authors recognised a small sample size and the presence of potentially relevant interfering factors. The study consisted of a correlation study of 36-month-old children between data obtained from the Mental Developmental Index and Psychomotor Developmental Index, while gut microbiota was assessed using 16S rRNA gene profiling. Children's gut microbiota explained a large portion of the variability in Bayley Scales, suggesting that gut microbiota may play an essential role in children's neurodevelopment.

During the gestation and postdelivery period, variations in maternal microbial populations have been suggested to modulate the microbiome, neurodevelopment, and behaviour of the offspring that express sexually dimorphic microbiome in neurological function (Jašarević et al. 2015, 2016). This is consistent with recent metagenomic studies showing that the microbiome exerts steroid hormone synthesising capacity (Hollister et al. 2015). These authors disagreed with previous studies suggesting that the human gut microbiome is relatively stable and adult-like after the first one to three years of life. According to their results, a healthy paediatric gut microbiome would harbour different compositional and functional qualities compared to healthy adults. Furthermore, they support the concept that the gut microbiome may undergo a more prolonged developmental stage than previously suspected.

Additionally, the following excerpts underline the complexity in analysing the microbiome-host relationship,

> The Human Microbiome Project. Little is known about the communities of microbial cells that inhabit healthy human bodies. The Human Microbiome Project aims to study these cells and their role in human health and disease. New DNA sequencing technologies have created a field of research, called metagenomics, that allows the comprehensive study of microbial communities, even those composed of organisms that cannot be cultivated experimentally. In the human micro-biome project, this method will complement genetic analyses of known isolated strains, providing unprecedented information about the complexity of human microbial communities.
>
> *(Kantor, NIH Roadmap for Medical Research 2008)*

> The gut microbiome is increasingly recognised as having fundamental roles in multiple aspects of human physiology and health including obesity, non-alcoholic fatty liver disease, inflammatory diseases, cancer, metabolic diseases, aging, and neurodegenerative disorders.
>
> *(Rothschild et al. 2018)*

These significant considerations should be considered when analysing pregnancy and delivery under different socio-economic conditions, most notably in its impact on infantile neuro-behavioural developmental conditions, as reviewed in McMath et al. (2023).

*

The incidence of the microbiome in mental health diseases has been considered by several authors, including Münger et al. (2018) and Shoubridge et al. (2022). They considered that the gut microbiome could influence neurophysiology and CNS function through several pathways, including the microbial production of short-chain fatty acids, vagus nerve stimulation, tryptophan production, and triggering of cytokine release. According to these authors, there is an increasing acceptance that the gut microbiome represents a considerable influence on brain physiology and mental health outcomes. Within the context of environmental factors, such as food or probiotics intake on brain activity, Allen et al. (2016) reported that consumption of psychobiotics affected in rodents the stress response, cognition, and brain activity patterns, a response in which vagal activity would be involved. Furthermore, according to Ribeiro et al. (2022), recent systematic reviews and meta-analyses have supported the role of the gut microbiome in mental health. In this domain, according to Allen et al. (2016), consumption of *Bifidobacterium longum* 1714™ in healthy volunteers (N= 22) is associated with reduced stress and improved memory. Furthermore, the effects of *B. longum* 1714™ on neural responses to social stress induced by the "Cyberball game," a standardised social stress paradigm, were studied by Wang et al. (2019) in humans using magnetoencephalography. According to the authors *B. longum* 1714™ modulated resting neural activity that correlated with enhanced vitality and reduced mental fatigue.

Li et al. (2008) observed a correlation between dietary-induced shifts in bacteria diversity and animal behaviour that may indicate a role for gut bacterial diversity in memory and learning. Tillisch et al. (2013) reported that consumption by healthy women of a fermented milk product with a probiotic was associated with changes in the response to acquired evoked and resting-state brain responses. This was tested using functional magnetic resonance imaging of a widely distributed brain network following exposure to a validated task probing attention to a negative context. This involved brain activity changes in several central neural circuits that the authors related to either vagal afferent changes or systemic metabolic changes. Buffington et al. (2016) reported that a marked shift in microbial ecology caused by a maternal high-fat diet can negatively impact social behaviours and related neuronal changes in offspring. Davison et al. (2018) considered that experimental examination of the microbiome through manipulation of diet, infection, stress, and exercise suggests direct effects on cognitive domains, including learning and memory. On this general domain, Fernandez-Real et al. (2015) considered that the human gut microbiota profile is significantly associated with brain microstructure (glial cells in mice) and cognitive function observed in middle-aged subjects before age-related cognitive decline (see also Meyer et al. 2022). Liang et al. (2022) remarked that the instability of gut microbial composition is correlated with cognitive impairment and stated that significant differences in gut microbial composition were observed among people with different cognitive statuses.

As posed by Parsons et al. (2020), there is increasing evidence that the microbiome affects neurodevelopment, but mechanistic causes are largely unknown. In this domain, Tamana et al. (2021) reported strong evidence of positive associations between *Bacteroidetes* gut microbiota in late infancy and subsequent neurodevelopment in human male infants using the Bayley Scale of Infant Development (BSID-III) at one and two years of age. Though the observed sexual dimorphic response would require further analysis, provisionally, its basis has been associated with increased male susceptibility to disruptions in the gut microbiome, as proposed by Jašarević et al. (2016). These authors further considered that since males and females exhibit sexually dimorphic patterns in energy and nutritional requirements across the lifespan, sex differences in the gut microbiome–brain axis may be an intervening factor in these processes. This would depend on sex-specific shifts in the ecological structure of the gut microbiome to meet different nutritional and energetic demands of growth, development, and reproduction. Additionally, Tamana et al. (2021) linked this sexual dimorphic response to early life differences in the microbiome gut-brain axis regulating the hippocampal serotonin system, as earlier reported by Clarke et al. (2013).

Regarding the reciprocal interactions between host and microbiome on social grounds, Münger et al. (2018) considered that throughout a long history of co-evolution, the gut microbiome may have become involved in the modulation of their host's sociality, resulting in fostering their transmission, with gut microbiome playing an important role in shaping host sociality. Thus, microbial life may have roles in multiple physiological processes, including mental health and behaviour in social settings. This has opened the possibility of applying a "gestalt perspective," allowing us to understand physiological, behavioural, and cognitive processes as part of an integrated whole. In this context, social behaviours observed between individuals with strong social bonds may allow the horizontal transmission of the gut microbiome. The authors further considered that gregariousness, social structures, and social behaviours might have partly evolved associated with the beneficial transmission of endosymbionts. Moeller et al. (2018) analysed the form of bacterial transmission affecting the gut microbiota in mice, concluding that horizontal transmission in mice and humans is more likely to exhibit virulence than vertically transmitted bacterial genera. However, most of the murine gut microbiota was vertically inherited.

Thus, previous comments stress the interaction between early microbiome establishment with developmental patterns in the neurocognitive and behavioural domains.

*

On pathologic grounds, the imbalance between the gut microbiome and the host is known for inducing diverse physiological deficiencies and pathological conditions. As discussed in Rautava and Walker (2007), commensal microbes and the host immune system live in a dynamic state of equilibrium. When disrupted, it contributes to the pathogenesis of intestinal inflammatory conditions (such as inflammatory bowel disease and necrotising enterocolitis). According to these authors, intestinal bacteria also contribute to healthy morphologic and functional maturation of the intestinal immune system in the postnatal period. Additionally, alterations in gut microbiota composition in infancy have been associated with subsequent development of immune-mediated disease. The microbiome encodes metabolic functions that humans have not had to evolve wholly on their own, including the ability to extract energy and nutrients from the diet, according to Ley et al. (2008). In this regard, Stappenbeck and Virgin (2016) reported studies that support the idea that the microbiome and metagenome (the sum of all host genes plus all genes of the microbiome) have profound local as well as systemic effects that would eventually include a wide range of potential pathologic conditions.

In concluding these observations, it seems opportune to mention that the 2012 report of the Human Microbiome Project Consortium made the following statement regarding the healthy human microbiome,

> Studies of the human microbiome have revealed that even healthy individuals differ remarkably in the microbes that occupy habitats such as the gut, skin and vagina. Much of this diversity remains unexplained, although diet, environment, host genetics and early microbial exposure have all been implicated. Accordingly, to characterise the ecology of human-associated microbial communities, the

Human Microbiome Project has analysed the largest cohort and set of distinct, clinically relevant body habitats so far. We found the diversity and abundance of each habitat's signature microbes to vary widely even among healthy subjects, with strong niche specialisation both within and among individuals. The project encountered an estimated 81–99% of the genera, enzyme families and community configurations occupied by the healthy Western microbiome.

It further stated that the extensive sampling performed of the human microbiome across many subjects and body habitats provides an initial characterisation of the normal microbiota of healthy adults in a Western population. The report asserts that the absence of particularly detrimental microbes supports the hypothesis that even given this cohort's high diversity, the microbiota tends to occupy a range of configurations in health distinct from many of the disease perturbations studied to date.

References

Allen, Andrew P., *et al*. "*Bifidobacterium Longum* 1714 as a Translational Psychobiotic: Modulation of Stress, Electrophysiology and Neurocognition in Healthy Volunteers." *Translational Psychiatry*, vol. 6, no. 11, 2016, doi:10.1038/tp.2016.191.

Braniste, Viorica, *et al*. "The Gut Microbiota Influences Blood-Brain Barrier Permeability in Mice." *Science Translational Medicine*, vol. 6, no. 263, 2014, doi:10.1126/scitranslmed.3009759.

Buffington, Shelly A., *et al*. "Microbial Reconstitution Reverses Maternal Diet-Induced Social and Synaptic Deficits in Offspring." *Cell*, vol. 165, no. 7, 2016, pp. 1762–1775, doi:10.1016/j.cell.2016.06.001.

Clarke, Gerard, *et al*. "The Microbiome-Gut-Brain Axis During Early Life Regulates the Hippocampal Serotonergic System in a Sex-Dependent Manner." *Molecular Psychiatry*, vol. 18, no. 6, 2013, pp. 666–673. doi:10.1038/mp.2012.77.

Davidson, Gabrielle L., *et al*. "The Gut Microbiome as a Driver of Individual Variation in Cognition and Functional Behaviour." *Philosophical Transactions of the Royal Society B: Biological Sciences*, vol. 373, no. 1756, 2018, p. 2017.0286, doi:10.1098/rstb.2017.0286.

Dinan, Timothy G., *et al*. "Collective Unconscious: How Gut Microbes Shape Human Behavior." *Journal of Psychiatric Research*, vol. 63, 2015, pp. 1–9, doi:10.1016/j.jpsychires.2015.02.021.

Dinan, Timothy G., and John F. Cryan. "Mood by Microbe: Towards Clinical Translation." *Genome Medicine*, vol. 8, no. 36, 2016, www.ncbi.nlm.nih.gov/pmc/articles/PMC4822287/, doi:10.1186/s13073-016-0292-1.

Dinan, Timothy G., and John F. Cryan. "Gut Instincts: Microbiota as a Key Regulator of Brain Development, Ageing and Neurodegeneration." *The Journal of Physiology*, vol. 595, no. 2, 2017, pp. 489–503. doi:10.1113/JP273106.

Dunbar, Robin I. M. "The Social Brain Hypothesis." *Evolutionary Anthropology: Issues, News, and Reviews*, vol. 6, no. 5, 1998, pp. 178–190, doi:10.1002/(SICI)1520-6505(1998)6:5%3C178::AID-EVAN5%3E3.0.CO;2-8.

Dunbar, Robin I. M. "The Social Brain: Mind, Language, and Society in Evolutionary Perspective." *Annual Review of Anthropology*, vol. 32, no. 1, 2003, pp. 163–181, doi:10.1146/annurev.anthro.32.061002.093158.

Dunbar, Robin I. M. "The Anatomy of Friendship." *Trends in Cognitive Sciences*, vol. 22, no. 1, 2018, pp. 32–51, www.sciencedirect.com/science/article/pii/S1364661317302243, doi:10.1016/j.tics.2017.10.004.

Fernandez-Real, José-Manuel, *et al*. "Gut Microbiota Interacts with Brain Microstructure and Function." *The Journal of Clinical Endocrinology & Metabolism*, vol. 100, no. 12, 2015, pp. 4505–4513, doi:10.1210/jc.2015-3076.

Gilbert, Scott F., *et al.* "A Symbiotic View of Life: We Have Never Been Individuals." *The Quarterly Review of Biology*, vol. 87, no. 4, 2012, pp. 325–341, doi:10.1086/668166.

Hair, Nicole L., *et al.* "Association of Child Poverty, Brain Development, and Academic Achievement." *JAMA Pediatrics*, vol. 169, no. 9, 2015, pp. 822–829, jamanetwork.com/journals/jamapediatrics/article-abstract/2381542, doi:10.1001/jamapediatrics.2015.1475.

Heijtz, R. D., *et al.* "Normal Gut Microbiota Modulates Brain Development and Behavior." *Proceedings of the National Academy of Sciences*, vol. 108, no. 7, 2011, pp. 3047–3052, doi:10.1073/pnas.1010529108.

Hoban, A. E., *et al.* "Regulation of Prefrontal Cortex Myelination by the Microbiota." *Translational Psychiatry*, vol. 6, no. 4, 2016, pp. e774–e774, www.nature.com/articles/tp201642, doi:10.1038/tp.2016.42.

Hollister, Emily B., *et al.* "Structure and Function of the Healthy Pre-Adolescent Pediatric Gut Microbiome." *Microbiome*, vol. 3, no. 36, 2015, pubmed.ncbi.nlm.nih.gov/26306392/, doi:10.1186/s40168-015-0101-x.

Jašarević, Eldin, *et al.* "Alterations in the Vaginal Microbiome by Maternal Stress are Associated with Metabolic Reprogramming of the Offspring Gut and Brain." *Endocrinology*, vol. 156, no. 9, 2015, pp. 3265–3276, doi:10.1210/en.2015-1177.

Jašarević, Eldin, *et al.* "Sex Differences in the Gut Microbiome–Brain Axis across the Lifespan." *Philosophical Transactions of the Royal Society B: Biological Sciences*, vol. 371, no. 1688, 2016, doi:10.1098/rstb.2015.0122.

Kantor, Lori Wolfgang. "NIH Roadmap for Medical Research." *Alcohol Research & Health*, vol. 31, no. 1, 2008, pp. 12–13.

Laue, Hannah E., *et al.* "The Developing Microbiome from Birth to 3 Years: The Gut-Brain Axis and Neurodevelopmental Outcomes." *Frontiers in Pediatrics*, vol. 10, 2022, doi:10.3389/fped.2022.815885.

Ley, Ruth E., *et al.* "Evolution of Mammals and Their Gut Microbes." *Science*, vol. 320, no. 5883, 2008, pp. 1647–1651, doi:10.1126/science.1155725.

Li, M., *et al.* "Symbiotic Gut Microbes Modulate Human Metabolic Phenotypes." *Proceedings of the National Academy of Sciences*, vol. 105, no. 6, 2008, pp. 2117–2122, doi:10.1073/pnas.0712038105.

Liang, Xinxiu, *et al.* "Gut Microbiome, Cognitive Function and Brain Structure: A Multi-Omics Integration Analysis." *Translational Neurodegeneration*, vol. 11, no. 49, 2022, www.ncbi.nlm.nih.gov/pmc/articles/PMC9661756/, doi:10.1186/s40035-022-00323-z.

Luczynski, Pauline, *et al.* "Adult Microbiota-Deficient Mice Have Distinct Dendritic Morphological Changes: Differential Effects in the Amygdala and Hippocampus." *European Journal of Neuroscience*, vol. 44, no. 9, 2016, pp. 2654–2666, doi:10.1111/ejn.13291.

McMath, Arden, *et al.* "A Systematic Review on the Impact of Gastrointestinal Microbiota Composition and Function on Cognition in Healthy Infants and Children." *Frontiers in Neuroscience*, vol. 17, 2023, doi:10.3389/fnins.2023.1171970.

Meyer, Katie, *et al.* "Association of the Gut Microbiota with Cognitive Function in Midlife." *JAMA Network Open*, vol. 5, no. 2, 2022, p. e2143941, doi:10.1001/jamanetworkopen.2021.43941.

Moeller, Andrew H., *et al.* "Transmission Modes of the Mammalian Gut Microbiota." *Science*, vol. 362, no. 6413, 2018, pp. 453–457, doi:10.1126/science.aat7164.

Morais, Livia H. *et al.* "The Gut Microbiota-Brain Axis in Behaviour and Brain Disorders." *Nature Reviews Microbiology*, vol. 19, no. 4, 2021, pp. 241–255, doi:10.1038/s41579-020-00460-0.

Münger, Emmanuelle, *et al.* "Reciprocal Interactions between Gut Microbiota and Host Social Behavior." *Frontiers in Integrative Neuroscience*, vol. 12, 2018, doi:10.3389/fnint.2018.00021.

Novotný, Miroslav, *et al.* "Microbiome and Cognitive Impairment: Can Any Diets Influence Learning Processes in a Positive Way?" *Frontiers in Aging Neuroscience*, vol. 11, 2019, doi:10.3389/fnagi.2019.00170.

Oluwagbemigun, Kolade, *et al.* "A Prospective Investigation into the Association Between the Gut Microbiome Composition and Cognitive Performance Among Healthy Young Adults." *Gut Pathogens*, vol. 14, no. 15, 2022, doi:10.1186/s13099-022-00487-z.

Parsons, Emilee, *et al.* "The Infant Microbiome and Implications for Central Nervous System Development." *Progress in Molecular Biology and Translational Science*, vol. 171, 2020, pp. 1–13, pubmed.ncbi.nlm.nih.gov/32475519, doi:10.1016/bs.pmbts.2020.04.007.

Penders, John, *et al.* "Factors Influencing the Composition of the Intestinal Microbiota in Early Infancy." *Pediatrics*, vol. 118, no. 2, 2006, pp. 511–521, www.ncbi.nlm.nih.gov/pubmed/16882802, doi:10.1542/peds.2005-2824.

Rautava, Ramuli, and W. Allan Walker. "Commensal Bacteria and Epithelial Cross Talk in the Developing Intestine." *Current Gastroenterology Reports*, vol. 9, no. 5, 2007, pp. 385–392, doi:10.1007/s11894-007-0047-7.

Ribeiro, Gabriela, *et al.* "Diet and the Microbiota–Gut–Brain-Axis: A Primer for Clinical Nutrition." *Current Opinion in Clinical Nutrition & Metabolic Care*, vol. 25, no. 6, 2022, pp. 443–450, doi:10.1097/mco.0000000000000874.

Roswall, Josefine, *et al.* "Developmental Trajectory of the Healthy Human Gut Microbiota during the First 5 Years of Life." *Cell Host & Microbe*, vol. 29, no. 5, 2021, pp. 765–776.e3, www.sciencedirect.com/science/article/pii/S1931312821001001, doi:10.1016/j.chom.2021.02.021.

Rothenberg, Sarah E., *et al.* "Neurodevelopment Correlates with Gut Microbiota in a Cross-Sectional Analysis of Children at 3 Years of Age in Rural China." *Scientific Reports*, vol. 11, no. 7384, 2021, doi:10.1038/s41598-021-86761-7.

Rothschild, Daphna, *et al.* "Environment Dominates over Host Genetics in Shaping Human Gut Microbiota." *Nature*, vol. 555, no. 7695, 2018, pp. 210–215. doi:10.1038/nature25973.

Shoubridge, Andrew P., *et al.* "The Gut Microbiome and Mental Health: Advances in Research and Emerging Priorities." *Molecular Psychiatry*, vol. 27, 2022, pp. 1908–1919, doi:10.1038/s41380-022-01479-w.

Smith, Leigh K., and Emily F. Wissel. "Microbes and the Mind: How Bacteria Shape Affect, Neurological Processes, Cognition, Social Relationships, Development, and Pathology." *Perspectives on Psychological Science*, vol. 14, no. 3, 2019, pp. 397–418, doi:10.1177/1745691618809379.

Stappenbeck, Thaddeus S., and Herbert W. Virgin. "Accounting for Reciprocal Host–Microbiome Interactions in Experimental Science." *Nature*, vol. 534, no. 7606, 2016, pp. 191–199, doi:10.1038/nature18285.

Stilling, Roman M., *et al.* "Friends with Social Benefits: Host-Microbe Interactions as a Driver of Brain Evolution and Development?" *Frontiers in Cellular and Infection Microbiology*, vol. 4, no. 147, 2014, doi:10.3389/fcimb.2014.00147.

Tamana, Sukhpreet K., *et al.* "Bacteroides-Dominant Gut Microbiome of Late Infancy Is Associated with Enhanced Neurodevelopment." *Gut Microbes*, vol. 13, no. 1, 2021, doi:10.1080/19490976.2021.1930875.

The Human Microbiome Project Consortium. "Structure, Function and Diversity of the Healthy Human Microbiome." *Nature*, vol. 486, no. 7402, 2012, pp. 207–214, doi:10.1038/nature11234.

Tillisch, Kirsten, *et al.* "Consumption of Fermented Milk Product with Probiotic Modulates Brain Activity." *Gastroenterology*, vol. 144, no. 7, 2013, pp. 1394–1401.e4, www.ncbi.nlm.nih.gov/pmc/articles/PMC3839572/, doi:10.1053/j.gastro.2013.02.043.

Tooley, Katie Louise. "Effects of the Human Gut Microbiota on Cognitive Performance, Brain Structure and Function: A Narrative Review." *Nutrients*, vol. 12, no. 10, 2020, p. 3009, doi:10.3390/nu12103009.

Tung, Jenny, *et al.* "Social Networks Predict Gut Microbiome Composition in Wild Baboons." *ELife*, vol. 4, 2015, doi:10.7554/elife.05224.

Villa, Tomás G., and Alicia Sánchez-Pérez. "The Gut Microbiome Affects Human Mood and Behavior." In *Developmental Biology in Prokaryotes and Lower Eukaryotes*, edited by Tomás G. Villa and Trinidad Bouzas. Cham: Springer, 2021, pp. 541–565, doi:10.1007/978-3-030-77595-7_22.

Wang, Huiying, *et al.* "*Bifidobacterium Longum* 1714[TM] Strain Modulates Brain Activity of Healthy Volunteers during Social Stress." *The American Journal of Gastroenterology*, vol. 114, no. 7, 2019, pp. 1152–1162.

8

THE MICROBIOME AND ITS BEHAVIOURAL IMPACT

We humans live in a microbe-dominated planet and have benefited from the "invention," early in metazoan evolution, of a gut that harbors microbial resources with the capacity to adaptively support metabolic activities not represented in the host genome.

(Goyal et al. 2015)

It is clear now that virtually no aspect of mammalian physiology and behaviour is left untouched by the community of microorganisms resident in our gastro-intestinal tract and the resultant host–microbiome interactions are critical for typical brain function and development.

(Clarke and Cryan 2016)

Previous chapters dwelled on the concept that life develops on two web-like para-digms. One is at the individual level, stemming from the ancestral microorganism domains where personal biological and social history contributes to generating indivi-dual profiles in interaction with the gut microbiome shared by familiar or cultural groups. The second one is on an extra personal dimension, a web that encloses humans as a subset of an interactive environmental concept, sharing members of Archean, prokaryotic, and eukaryotic organisms within a conditional universe of living creatures. Gut microbiota per person has been estimated by Frank and Pace (2008) to consist of at least 1800 genera, up to 40,000 species of bacteria, have an estimated mass of one to two kg, and number 100 trillion. As reported by Kurokawa et al. (2007), together, these possess 100 times the number of genes in the human genome, though challenged by Sender et al. (2016) to be in the order of 1:1.

On pragmatic grounds, what is the evidence of microbiomes affecting human beha-viour and health? Brown et al. (2021) called attention to a large group of genes within the human microbiome that code for transferases predicted to manipulate host cells.

DOI: 10.4324/9781032698380-8

By influencing individual interaction and genetic transfer, microbiota could be a critical component in the evolution of metazoan species (Sampson and Mazmanian 2015). Growing evidence suggests that the mammalian microbiome can affect behaviour and that several symbionts produce neurotransmitters. In this regard, Johnson and Foster (2018) considered that microbiome behavioural effects can readily arise as a by-product of natural selection on microorganisms grown within the host. The authors argued that understanding microbiome influences on behaviour requires focusing on microbial ecology and local effects within the host.

As mentioned in previous chapters, metazoans evolved in a world dominated by microbial life. During evolution, the colonisation of gut microbiota in mammalian species became integrated into the programming of brain development, affecting motor control and anxiety/arousal-like behaviour. Thus, microbiota is increasingly recognised for its ability to influence the development and function of the nervous system and several complex host behaviours. Within the pathological domain, the cases of rabies virus in humans and parasitic microbes in insect behaviour, as well as experiments in mice, impact host neurotransmitters (Bravo et al. 2011; Foster and McVey Neufeld 2013; Hsiao et al. 2013; Desbonnet et al. 2014; Yano et al. 2015) and mind/behavioural alterations (Lyte 2013; Cryan and Dinan 2012). These are only a few examples of reported correlations of microbiome influence on behaviour. On experimental grounds, germ-free mice have been critical in assessing the role of microbiota in various aspects of physiology, e.g., displaying deficits in simple non-spatial and working memory tasks (Gareau et al. 2011). Sarkar et al. (2020) added that,

> Microbes colonise all multicellular life, and the gut microbiome has been shown to influence a range of host physiological and behavioural phenotypes. One of the most intriguing and least understood of these influences lies in the domain of the microbiome's interactions with host social behaviour, with new evidence revealing that the gut microbiome makes important contributions to animal sociality.

According to Vuong et al. (2017), perturbations in the microbiota have been associated with changes in social, communicative, stress-related, and cognitive behaviour in laboratory and wild animals. Findings suggest that specific bacterial species from the gut microbiota can influence social-communicative behaviour in a postnatally inducible and reversible manner. The former authors cite that select probiotic treatments can modulate animal learning and memory behaviour. As proposed by Sampson and Mazmanian (2015), the observed host-microbiome correlations on behavioural domains may not necessarily be a direct function of specific species bacterial components. However, they would imply a broader role of the community of symbiotic bacteria within the gut. If so, the beneficial effects of single probiotics on some morbidities may suggest a reformulation of gut microbiome interactions.

*

According to Stilling et al. (2014), enhanced microbial transmission through group living and expanded non-coding RNA regulation would have contributed to advanced social behaviour in primates and human intelligence. Furthermore, social complexity would increase the transmission of beneficial microbes, causing host-symbiont coevolution on RNA regulation in the brain. The impact of social behaviour on host-microbiome expansion has also been considered by these authors. Stilling et al. (2014) support a model in which the evolution of human sociability, which was accompanied by an accelerated extension of the neocortex, is a key example of host-microbe co-evolution and would depend on endosymbiotic developmental signals through the microbiota-gut-brain axis. Interestingly, following the analysis of the impact of social behaviour on the microbiome in chimpanzees, Moeller et al. (2016) commented that chimpanzee social interactions propagate microbial diversity in the gut microbiome within and between host generations and further added that,

> ... results indicate that social behaviour generates a pan-microbiome, preserving microbial diversity across evolutionary time scales and contributing to the evolution of host species–specific gut microbial communities.

Jašarević et al. (2016) added a conceptual framework for sexually dimorphic communication between the gut microbiome and the brain, stressing the potentially different energy and nutritional demands of each sex. This would suggest sex-specific shifts in the ecological structure of the gut microbiome to meet these demands and the critical involvement of a sexually dimorphic microbiome in neurological function and dysfunction.

The behavioural impact of host-microbiome interactions has been further tackled by Johnson and Foster (2018). They provided a series of considerations centred on the premise that behavioural effects can readily arise as a by-product of natural selection on microorganisms to grow within the host. Additionally, natural selection on hosts would depend upon their symbionts. The authors further proposed that understanding why the microbiome influences behaviour requires focusing on microbial ecology and local effects within the host. These authors considered manipulation by the microbiome of host behaviour is often unlikely, though they bring up a different perspective on the host impact of microbiome interactions by exploring other evolutionary explanations for the behavioural effects of mammalian symbionts. According to these authors, the physiological construction of such host-microbiome interaction could be summarised in terms of whether hosts evolve to depend on microbial metabolites for normal physiological function, with behavioural dysfunction emerging otherwise. In conclusion, according to Johnson and Foster (2018),

> If we have evolved to depend on members of the microbiota to modulate our own neurochemistry, then we might expect their absence to influence brain function. Thus, hosts are expected to evolve to depend upon, monitor and regulate their microbiota. This evolution may readily forge and modulate links between the microbiota and host behaviour... We should not then assume that our microorganisms are our puppeteers. Instead, the behavioural effects of the microbiota might be better explained as a side effect of either local manipulation of the host environment or the microbial metabolism needed to grow and survive in the gut.

As discussed in Hanstock et al. (2004) and Montiel-Castro et al. (2013), studies on the effects of intestinal microbiota composition on brain function predominantly involved animal models of behavioural disorders such as anxiety, depression, and cognitive dysfunction. The authors considered the ability of the gut microbiota to communicate with the brain and thus modulate behaviour and affect the ethological and cultural strategies of human and nonhuman primates to select, transfer, and eliminate microorganisms by selecting the commensal profile. Accumulating evidence suggests that the composition of the gut microbiota may also have a role in several other metabolic conditions that involve the CNS. In this regard, the gut microbiota has demonstrated unique functions associated with behaviour, mood, and cognition. For example, Mayer (2011) reported evidence suggesting that various forms of subliminal interoceptive inputs from the gut, including those generated by intestinal microbes, may influence memory formation, emotional arousal, and affective behaviours. This author considered the human insula and related brain networks to be the most plausible brain region to support this integration. Furthermore, according to this author, research performed on this gut-brain crosstalk has revealed a complex, bidirectional communication system likely to have multiple effects on affect, motivation, and higher cognitive functions. In this regard, shifting to experimental grounds, analysing memory disfunction related to altered gut microbiota, Gareau et al. (2011) reported that memory was impaired in germ-free mice, with or without exposure to stress, in contrast to control mice with intact intestinal microbiota. According to a later report by Gareau (2016), evidence for cognitive deficits has now been identified in numerous intestinal and extraintestinal diseases. Also, as mentioned by Bercik et al. (2012),

> The effect of microbiota may extend into memory and cognition, as recently suggested in a study comparing germ-free and SPF mice. An Earlier study by Li et al. examined the effects of long-term dietary manipulation on memory.

Forsythe et al. (2010) considered that evidence points towards a possible role for gastrointestinal commensal microbes in the modulation of nervous system function. Carrier and Reitzel (2017) support that the hologenome paradigm of animals and plants with their corresponding microbiomes serves as a unit of selection, as stated by Zilber-Rosenberg and Rosenberg (2008). The former authors further sustain that this paradigm has led to advancements in our understanding of the spectrum of organismal symbioses in the life sciences regarding the host's developmental, evolutionary, and genetic modifications. Furthermore, it supports the concept that partnerships between host and microbes within an environmental setting exemplify a network of biotic relationships common across the tree of life. However, this partnership would vary across the gut sections. In this domain, Lyte (2013) called attention to the fact that the innervation of the gut is not homogenous throughout its length. Neither is the microbiome. Hence, according to this author, it would be necessary to understand how one microbial species that may produce a neuroactive compound may have a behavioural effect while in one section of the gut and not another. Liang et al. (2023) reviewed the interaction of gut microbiota and brain function, stressing the impact of dietary compounds.

*

Systemic Impact

Given the prolonged natural history of microbiomes associated with hijacking plant and animal species, it should not surprise the various possible mechanisms and host consequences involved in this association. This has been functionally considered an organ of postnatal acquisition, and it performs a set of different functions for the host. Initial experiments involved animal models that provided evidence of possible neural and metabolic pathways affected by commensals and probiotics affecting visceral perception, as Verdu et al. (2006) reported. Experimental use of colorectal distension provided a design to test several probiotic bacteria for their ability to decrease pain perception during this procedure (Kamiya et al. 2006; McKernan et al. 2010).

As mentioned by Bercik et al. (2012) the gut microbiome would play a crucial role in developing innate and adaptive immune responses and influence physiological systems throughout life. This would be performed by modulating gut motility, intestinal barrier homeostasis, absorption of nutrients and the distribution of somatic and visceral fat. The former authors cite a series of experimental procedures to analyse microbial involvement in behavioural performances, including influence on host behavioural phenotype.

Regarding microbiome interaction with gut physiology, earlier studies performed comparisons of germ-free and colonised mice and revealed that microbes drive the production of mucosal immunoglobulin, as mentioned in Shroff et al. (1995). Previously, in 1986, Savage had commented that gastrointestinal microflora is known to serve nutritional functions in ruminants, pseudo ruminants (lacking a four-compartment stomach), and monogastric mammals with only modest or no foregut fermentations but with extensive hindgut fermentations in blind caecal pouches. More recently, Hooper et al. (2002), and Hooper (2004) concluded that the intestine is genetically pre-programmed to receive bacterial signals during weaning and that the postnatal mammalian intestine appears to be poised for interaction with its microbial partners, which are essential for its normal development. This would involve shifts in the glycoconjugate repertoire, microbicidal protein expression, and angiogenesis, critical events for postnatal gut development. Additionally, as reported by Hooper (2004), studies in germ-free mice have indicated that gut bacteria influence the maturation and function of several components of the mucosal immune system.

*

Microbiome Impact on Behaviour and Cognition

As commented earlier, the microbiome has a measurable impact on the brain, influencing stress, anxiety, depressive symptoms, and social behaviour. The microbiome–gut–brain axis may be mediated by various mechanisms, including neural, immune, and endocrine signalling and has been shown to be linked to personality traits in the general population, as mentioned by Johnson (2020).

On neuroendocrinological domains, Sudo et al. (2004), tested whether postnatal microbial colonisation would affect the development of neural systems (see also Parsons et al., 2020) that govern the endocrine stress response. Results disclosed that it depended on the age at which microbial colonisation took place. The authors further concluded that exposure to microbes at an early developmental stage was required for the hypothalamic–pituitary–adrenal system to become fully susceptible to inhibitory neural regulation. Later, Neufeld et al. (2011) examined behaviour in germ free mice and provided evidence that gut microbiota is important for the development of stress circuitry and related behaviours. The authors also included the concept of *critical developmental period* in these interactions.

On behavioural grounds, Felice and O'Mahony (2017), confirmed alterations of the gut microbiota associated with stress-related disorders including depression and anxiety and irritable bowel syndrome.

Bailey et al. (2011) considered that stressor exposure significantly changed the community structure of the microbiota, affecting the relative presence *Bacteroides* and *Clostridium* favouring the latter, though mechanisms involved are still speculative. It seems worth including an additional comment from these authors related with microbiota in the elderly, who stated that it would act as a modulator of inflammatory processes in the brain, underlying many age-associated neurological diseases.

As quoted by Eloe-Fedrosh and Rasko (2013), by the first year of life, the microbiota has dynamically progressed and converged toward a stable, phylogenetically diverse, adult-like profile according to distinct "environmental" events. Borre et al. (2014) supported the concept that bacterial colonisation of the gut is central to postnatal development and maturation of key systems that have the capacity to influence central nervous system programming and signalling, including the immune and endocrine systems. Coincidentally, Carabotti et al. (2015), quoted experimental studies on germ free animals showing that bacterial colonisation of the gut is central to the development and maturation of both enteric and central nervous system. The absence of microbial colonisation would be associated with an altered expression and turnover of neurotransmitters in both nervous systems.

Gacias et al. (2016) supported the concept that microbiota-driven transcriptional changes in the prefrontal cortex override genetic differences in social behaviour. They considered that their results provide strong evidence that manipulations of gut microbiota are sufficient to induce depressive-like behaviours in adult mice. In this regard, Allen et al. (2017) affirmed that gut microbiota has been shown to interact with host cognition in numerous laboratory animal model studies and quoted reports on germ-free animals displaying altered social behaviour. According to Lu et al. (2018), early microbiome colonisation patterns influence growth trajectories, suggesting colonisation of different microbial communities influences brain development. Therefore, the initial colonisation of microbiota can be a modifiable factor affecting brain development. The authors further described that mouse colonised with microbiota from a preterm infant with good growth had significantly more weight gain postnatally than those colonised with microbes from a preterm infant with poor postnatal growth.

Dinan et al. (2015) posed that postnatal gut microbial colonisation occurs parallel to cognitive development. According to these authors, increasing evidence supports the view that the evolving cognitive activity is critically dependent on the microbiota and its metabolic activity. Furthermore, they state that data point to a critical role of microbial cues that affect brain development, especially impacting neuronal circuitries that underlie social behaviour in mammals. On developmental grounds, Carlson et al. (2018) considered that microbial composition of the human gut at one year predicts cognitive performance at two years of age, particularly in communicative behaviour. These results may have implications for developmental progress or predisposition to disorders characterised by cognitive or language delay in human populations compromised due to socioeconomic conditions.

Animal models have provided evidence of possible neural and metabolic pathways that are affected by commensals and probiotics. One of the first reports on the role of bacteria on visceral perception was provided by experiments performed by Verdu et al. (2006) in which microbiota from healthy NIH Swiss mice were altered by administration of non-absorbable antibiotics.

Within the millions of neurons in the enteric nervous system, a percentage of them would be sensing and communicating changes to the brain channelled through the vagus nerve (Li et al. 2009). Bercik et al. (2012) summarised the "gut-brain" axis as a bi-directional communication system. It would include the enteric nervous system, vagus nerve, sympathetic and spinal nerves, and humoral pathways, involving cytokines, hormones, and neuropeptides as signalling molecules, as also explored and reviewed by Cani and Knauf (2016).

As mentioned by Sharon et al. (2016), dietary components would interact directly with the developing brain and induce functional alterations in the mature brain. Clinical studies have related imbalances in dietary intake of fatty acids to impaired brain performance and diseases, as discussed in Chang et al. (2009).

Li et al. (2009) compared rats fed with standard chow diet or chow containing 50% lean ground beef. Results demonstrated significantly higher bacterial diversity in the beef-supplemented diet group. Compared to the purine-based diet (purines are aromatic organic compounds), the beef diet fed mice displayed improved working and reference memory, and a correlation between dietary induced shifts in bacteria diversity and animal behaviour that suggests a role for gut bacterial diversity in memory and learning. According to Stilling et al. (2014), some microbial products can modulate the epigenetic landscape of the host brain, i.e., stable changes in cell function that do not involve alterations in the DNA sequence. This would include regulators of the activity of histone-modifying enzymes either through metabolic alterations or direct interactions between bacteria-secreted molecules and host signalling pathways, providing cues for host neurodevelopment. These authors add that given a potential co-evolution of social behaviour in mammals and their microbiomes; brain development would be particularly vulnerable to signals from the microbiome. In later studies, Louwies et al. (2020) reinforced the concept that the microorganisms that reside in the gut can exert an important influence on the central nervous system through the gut-brain axis. According to these authors, signals arising from the gut can trigger neurodevelopmental and neurobehavioural effects on the brain. During critical neurodevelopmental periods such as early life, gut microbial composition changes may detrimentally impact neurodevelopment. Furthermore, transferring the faecal microbiota from patients with irritable bowel syndrome (IBS) or major depressive disorder to germ-free (GF) animals also transferred IBS-like visceral pain and depressive-like behaviour.

Hence, current data support the concept that dysfunction of the brain-gut microbiome axis has been implicated in behavioural stress-related disorders (depression, anxiety, irritable bowel syndrome, and neurodevelopmental disorders such as autism).

*

The above comments are not germane to modern diets and their cultural variations that represent a significant variability in the gut microbiome. Garud et al. (2019) discussed data proposing that the population of different species and strains can vary dramatically based on diet, host species, and the identities of other co-colonising taxa. These rapid shifts in gut community composition suggest that individual gut microbes may be adapted to specific environmental conditions, with strong selection pressures between competing species or strains. As analysed by Walter and Ley (2011), modern diets even richer in simple substrates compared with Neolithic diets further stress the interactions between the microbiota and the human host.

Furthermore, interactions between human genetics, diet, and the microbiota fundamentally shaped modern populations. Interestingly, on this account, several recent studies have found that people in industrialised nations host far fewer types of microbes than hunter-gatherers in Africa, Peru, and Papua New Guinea. De Filippo et al. (2010) compared the faecal microbiota of 15 healthy European children living in the urban area of Florence, Italy, breastfed for up to one year, to those of 14 healthy children from a Mossi ethnic group in Burkina Faso who were breastfed up to two years where the diet, high in fibre content, is like that of early human settlements at the time of the birth of agriculture. Besides differences in breast feeding, the authors proposed that gut microbiota coevolved with a polysaccharide-rich diet, allowing them to maximise energy intake from fibres and protecting them from inflammations and non-infectious colonic diseases. In this regard, observations made in mice (De Vadder et al. 2014) point to the role of gut microbiota in the fermentation of soluble fibre, which generates short chains of fatty acids associated with body weight and glucose control.

*

Zhu et al. (2020) stressed that thus far, evidence has revealed associations between microbiota and host physiology rather than demonstrating causal relationships. Due to multiple confounding variables in human faecal experiments, large sample-size studies are needed for metagenomic biomarker screening, besides the mild and long-term nature of effects observed in terms of cognitive or psychological domains. Similar considerations were made by Angoorani et al. (2022) in their review, which included animal and human reports. Though they concluded that the gut microbiota may alter brain function or trigger various psychiatric conditions through the gut-brain axis, they stressed the limited clinical studies performed so far and the number of difficulties in extrapolating the results of animal models to humans.

According to Zhu et al. (2022a), due to the bacterial endosymbiotic nature of mitochondria, bacteria and mitochondria share bioactive compounds and analogue genomic characteristics. Thus, due to their homology, metabolites of the gut microbiome would remotely regulate the function of neuronal mitochondria of brain tissue, generating a crosstalk among them. According to these authors, symbiont and pathobiont bacteria have the possibility of influencing neuronal mitochondrial activity; additionally, according to these authors,

In recent years, a mass of research has identified that gut microbiota and corresponding bacterial metabolites can target the brain through various pathways, such as nervous conduction (enteric nerve, vagus nerve, etc.) (Fulling et al., 2019), hypothalamic–pituitary–adrenal (HPA) axis (McNeilly et al., 2010), and enteric endocrine and immune response (Fung, 2020; Morais et al., 2020).

On behavioural grounds, according to Lyte (2013), microorganisms contained within the microbiome can influence behaviour through neurochemicals analogous in structure to those produced by the host nervous system. This form of interkingdom signalling, which is based on bidirectional neurochemical interactions between the host's neurophysiological system and the microbiome, was introduced earlier and has been termed *microbial endocrinology* by Lyte (1993). As this author states, the presence of hormone-like compounds and their receptors in micro-organisms has been recognised for several years and is believed to represent a form of intercellular communication. As such, it may constitute a type of primitive neuroendocrine system.

Though the concept of a *microbiome organ* has been challenged, Lyte (2010) considers that,

> Based on the ability of bacteria to both recognize and synthesize neuroendocrine hormones, it is hypothesized that microbes within the intestinal tract comprise a community that interfaces with the mammalian nervous system that innervates the gastrointestinal tract to form a microbial organ.

Furthermore, (Lyte 2013) added that,

> The extent of the prevalence of neuroendocrine hormones in nature that are exactly the same in structure, and most interestingly biochemical synthesis pathways, is often not fully appreciated. For example, the neuroendocrine hormone norepinephrine is found in plants, as well as in insects and fish, and most critically from the standpoint of microbiologists, in microbes. Indeed, due to these same shared biochemical pathways, the existence of neurochemical-based cell-to-cell signaling pathways such as those in humans has been proposed to be due to late horizontal gene transfer from bacteria [Iyer et al., 2004].

In this regard, Iyer et al. (2004) stated earlier on evolutionary grounds that the history of most genes encoding enzymes involved in the metabolism of these messengers is best described by scenarios that include horizontal gene transfer from bacteria, with some transfers occurring after the divergence of animals from fungi. The acquisition of bacterial genes via horizontal gene transfer seems to have had the essential role of extending existing biochemical pathways.

*

Regarding its behavioural impact, according to Bercik et al. (2011), microbiota composition has been shown to modulate anxiety-like behaviours in adult mice via changes in levels of brain-derived neurotrophic factor in the hippocampus. As stated in

Collins et al. (2013), evidence of microbiome-brain interactions in mice shows the ability to transfer behavioural traits between mouse strains using faecal microbiota transplantation (FMT), which alters brain chemistry and behaviour in recipient germ-free mice. Johnson (2020) quoted experiments performed in mice, implementing faecal microbiota transplantation, demonstrating that the gut microbiome can influence stress response, anxiety, depressive-like behaviours, social behaviour, and communication. Also, within neuro-behavioural domains, D'Amato et al. (2020) reported the impact of faecal microbiota transplant from aged donor mice, leading to a decline of spatial learning and memory in young adult recipients. These experimental results suggest that gut microorganisms can contribute causally to behavioural traits and open the possibility of using FMT in central nervous system disorders.

These experiments have a human counterpart since, according to Kurokawa et al. (2018), similar faecal microbiota transplantation performed in humans modified gastrointestinal symptoms, depression, anxiety, and sleep, suggesting that it might be effective in modifying psychiatric symptoms. These observations were extended by Tengeler et al. (2020) on experimental grounds. These authors reported that transplant to mice of bacterial components of the gut microbiota from individuals with attention-deficit/hyperactivity disorder, are associated with changes in mice brain structure and function and behaviour. Additional evidence of host-microbiome interactions on behavioural grounds was provided by experimental results following the effect of probiotics on altered social behaviour due to diet imbalances (Buffington et al. 2016). Though host microbiota modulates the development of social preference in mice (Arentsen et al. 2015), it would depend on host nutritional choices, which can modify the relative abundance of microbe types in their gut (Pasquaretta et al. 2018).

According to Sarkar et al. (2020), in addition to modulating brain physiology, the microbiome may also affect the central nervous system via the generation and regulation of a range of *social-signalling molecules*, including concentration of glucocorticoids, sex steroids, neuropeptides, and monoamines. The authors expanded these actions to the possible modulation by the microbiome of gene activity relevant to sociality, though Sarkar et al. (2020) concluded that human microbiome research is typically underpowered, and a great deal of variation characterises the data. Therefore, studies investigating the relationship between the human gut microbiome and psychology occasionally report inconsistent findings.

On experimental comparative grounds, Bruckner et al. (2022) analysed the zebrafish as a model of microbiome-host interactions as related to social behaviour and how the microbiota influences the development of the "social brain". They concluded that results would demonstrate that the microbiota influences zebrafish social behaviour involving microglial remodelling of forebrain circuits during early neurodevelopment. Additionally, according to these authors, and considering their results that microbiota promotes forebrain microglial abundance and gene expression, their experiments suggest that a feature common to many bacterial taxa promotes social behaviour by stimulating host innate immune pathways that redistribute forebrain microglia, alter microglial function, and remodel neuronal connections. Also, host-associated microbes would influence social behaviour across taxa, correlating with microbial modulation of neuronal gene expression, neurotransmitter levels, brain maturation, and myelination. In this regard, Bruckner et al. (2022) concluded that distinct microbiota compositions could result in variable levels of microglial complement signalling involving synaptic pruning, predisposing some individuals to neurodevelopmental disorders such as autism spectrum disorder.

Mayer et al. (2014) considered that the role of intact gut microbiota in shaping brain neurochemistry and emotional behaviour, primarily based on experiments performed in mice, has given way to an unprecedented potential paradigm shift in conceptualising many psychiatric and neurological diseases. However, it is currently unclear what the translational value of the results obtained in rodent models for understanding brain-gut disorders in humans will be. This issue was considered by Stilling et al (2015), who stated that though findings were made in experimental animals, there are data from healthy subjects and patients that generally support strong links between the microbiota in normal brain functioning and human neuro-behavioural outcomes. This was reported by Messaoudi et al. (2011) based on the effects of probiotics on anxiety and depression-related behaviours in human volunteers; by Tillish et al. (2013) reporting on the effects of fermented milk with probiotics in women and on brain activity involved in central processing of emotion; and by Allen et al. (2016), who reported that consumption of *Bifidobacterium longum* 1714 is associated with reduced stress and improved memory.

*

Foster et al. (2015) reported that the range of neurohormones found in microorganisms is extremely diverse, ranging from somatostatin to acetylcholine to progesterone. Critically, microorganisms that inhabit the gastrointestinal tract can produce neurochemicals that can bind to host receptors sufficiently to induce neurophysiological changes in the host.

> Production and recognition of neurochemicals that are more commonly associated with mammals by prokaryotic and eukaryotic microorganisms have led to a new understanding of an evolutionary-based mechanism by which microbes can influence host behaviour.
>
> *(from review by Foster et al. 2015)*

Gacias et al. (2016) identified gut microbiota as capable of inducing depressive-like behaviours and affecting social behaviour in genetically distinct mouse strains. This would occur partly by gut microbiota modifying the synthesis of key metabolites affecting gene expression in the prefrontal cortex. In this regard, alterations of the microbial composition modified gut-produced metabolites and transcriptomic profiles in the medial prefrontal cortex, subsequently affecting behaviour.

Regarding non-human primates, social contact is another mechanism that can mediate the acquisition and exchange of microbial symbionts, as Ezenwa et al. (2012) commented. On these grounds, Degnan et al. (2012) observed that chimpanzees from the same community have more similar microbial consortia than do chimpanzees from different communities and concluded that,

Despite the broad similarity of community members, as would be expected from shared diet or interactions, long-term immigrants to a community often harbored the most distinctive gut microbiota, suggesting that individuals retain hallmarks of their previous gut microbial communities for extended periods. This pattern was reinforced in several chimpanzees sampled over long temporal scales, in which the major constituents of the gut microbiota were maintained for nearly a decade.

On these domains, Amato (2016) considered that the impact of gut microbiota on host behaviour could be an essential factor in primate ecology due to its influence on individual personality, social status, host aggression, and anxiety levels. As described in Ochman et al. (2010), there are suggestions stemming from observations on great apes' gut microbial communities concordant with the phylogeny of their host species, promoting the specificity and co-diversification of bacterial communities with hosts. Furthermore, Arumugam et al. (2011) described human enterotypes, which correspond to groups of individuals clustered in characteristic gut microbiota profiles. These are reminiscent of core groups of resident taxa in the Gombe chimpanzee, as described by Degnan et al. (2012). Additionally, these authors mentioned that geographic, temporal, sex- and age-specific factors are associated with the long-term composition and diversity of the gut microbial communities harboured by Gombe chimpanzees. Their communities, though, remain distinct from those of other great apes, including other subspecies of chimpanzees. In wild baboons, Tung et al. (2015) reported the impact of social behaviour on gut microbiome, stressing that social interactions are an important determinant of gut microbiome composition in natural animal populations. These authors supported the hypothesis that social interactions play a role in the health-related consequences of variation in gut microbiome composition with potentially important consequences for the evolution of sociality, as also mentioned in Lombardo (2007), Ezenwa et al. (2012), Montiel-Castro et al. (2013), Miller et al. (2016) and Münger et al. (2018). Conversely, the social structuring of the microbiota appears to be shared. As observed in comparative animal studies, with whom an animal interacts and what they do together can have profound consequences for the composition of their microbiota (Archie and Tung 2015).

On neurobiological grounds, research performed in preclinical models suggests that the microbiome's effect on behaviour (including emotional behaviour) may be related to changes in the amygdala and hippocampus (Mohajeri et al. 2018). In this respect, the authors quote results suggesting significant differences in the volume and dendritic morphology of the amygdala and hippocampus between conventionally colonised and germ-free mice. In fact, according to Sarkar et al. (2020), previous studies in germ-free mice (Luczynski et al. 2016) revealed that the microbiome affects the structure and function of the amygdala.

If the microbiome is affecting brain morphology and function and hormonal and neurotransmitter signalling, it can be expected to influence host gene expression (Sarkar et al. 2020). The authors further discuss that since the specific mechanisms by which the microbiome influences host physiology remain poorly understood, an important and currently unresolved question is the order of microbial effects on host social development and behaviour. Germ-free status has been linked to many physiological impairments in mice, supporting the claim that microbes are essential for normal development.

Brain developmental susceptibility to given noxa involves the juvenile period. This concept was implemented by Desbonnet et al. (2015) to analyse in rats the impact on adult behaviour of gut microbiota depletion induced by chronic antibiotic administration since early adolescence. This treatment resulted in cognitive deficits, altered dynamics of the tryptophan metabolic pathway, and significantly reduced brain-derived neurotrophic factor, oxytocin, and vasopressin expression in the adult brain. The reported data would suggest,

> … that despite the presence of a normal gut microbiome in early postnatal life, reduced abundance and diversity of the gut microbiota from weaning influences adult behaviours and key neuromodulators of the microbiota–gut–brain axis suggesting that dysregulation of this axis in the post-weaning period may contribute to the pathogenesis of disorders associated with altered anxiety and cognition.

Regarding serotonin involvement in brain-gut microbiome axis, O'Mahony et al. (2015) pointed to the microbial influence on tryptophan metabolism and the serotonergic system as an essential node in the central nervous system and the gastrointestinal tract regulation.

As stated by Ezenwa et al. (2012), experiments performed on mice showed that the gut microbiome can influence stress, anxiety, and depression-related behaviour via effects on the host's neuroendocrine system, a neuroactive observation that humans would share, according to Valles-Colomer et al. (2019). This provides insight into how information can be passed between the host and microbe (Bravo et al. 2011, Heijtz et al. 2011). Regarding the role of the microbiome in neural pathologies, as mentioned by Wang and Kasper (2014), neural, endocrine, and metabolic mechanisms are critical mediators of the microbiome-central nervous system signalling. They would be involved in neuropsychiatric disorders such as autism, depression, anxiety, and stress (see also Nikolova et al. 2021; McGuiness et al. 2022; Safadi et al. 2022 regarding other neuropathologies). Bauer et al. (2019) considered that the gut microbiome contributes to normal brain function and pathologies, including higher neurological functions. In humans, Johnson (2020) considered that although there is evidence of gut microbiota dysbiosis implications in neuro-behavioural pathologies and social development and behaviour, it remains to be established if gut microbiome variations are related to personality traits. According to Palepu et al. (2022), depression-induced changes by the commensal intestinal microbiota are primarily linked with the disruption of gut integrity. Zhu et al. (2022b) considered that the incidence of depression is closely linked to the gut microbiota. The transformation of the gut microbiota system's structure is considered to have positive and negative regulatory effects on the development of depression.

The gut-brain axis role has been identified in several physiological processes such as satiety, food intake, fat and bone metabolism, glucose regulation, and insulin secretion (Oroojzadeh et al., 2022). The authors concluded that consuming psychobiotics containing nutrients affects the improvement of microbiota. This coincides with alleviating some symptoms of central nervous system pathologies, including autism spectrum disorders, Parkinson's disease, multiple sclerosis, insomnia, depression, diabetic neuropathy, and anorexia nervosa.

From the above reports, it becomes apparent that host-microbiome interactions could be involved in altered patterns of behaviours with personal and social impact, starting during neurodevelopmental stages and projected into social behavioural components in later life. This issue should be included in studies regarding current models of individual and collective social behaviours linked to socioeconomic conditions and the impact of nutrients availability and nutritional culture.

*

Neurocognitive and Social Behavioural Impact

According to Shoubridge et al. (2022), it remains challenging whether specific gut microbiome features contribute causally to neurological, psychological, or mental health conditions. However, several reports have disclosed its close relationship with behavioural and neuropathological disorders. They include Huntington's disease and multiple sclerosis; schizophrenia and major depressive disorder; and behavioural conditions, including autism spectrum disorder (see also Chen et al. 2021) and attention-deficit/hyperactivity disorder. These observations prompted increasing acceptance that the gut microbiome greatly influences brain physiology and mental health outcomes. Several possible routes including bidirectional communication would mediate this. Those routes would include microbial production of short-chain fatty acids, vagus nerve stimulation, tryptophan production, triggering cytokine release, the enteric nervous system, sympathetic and spinal nerves, and humoral pathways, which include cytokines, hormones, and neuropeptides as signalling molecules (see Furness, 2012; Sharon et al. 2016; Suganya and Koo 2020, for a description of the enteric nervous system and neuroendocrine central connections). According to Suganya and Koo (2020), the gut microbiota could communicate with the central and enteric nervous systems by producing several metabolites/neurotransmitters with neuro-modulatory properties.

Liang et al. (2022) reported significant differences in the gut microbial composition among people with different cognitive conditions after performing cognitive assessment applying the Mini Mental State Examination. They additionally revealed that increased individual alterations in gut microbial composition were associated with cognitive decline. The authors identified three genera – *Odoribacter, Butyricimonas*, and *Bacteroides* – which were depleted in participants with cognitive impairment compared to normal controls. These studies revealed that specific gut microbial features are closely associated with cognitive impairment and decreased hippocampal volume and that increased intraindividual alterations in gut microbial composition were associated with cognitive decline. A review of the impact of probiotics on cognitive decline was provided by Baldi et al. (2021). In clinical domains, the microbiome has been implicated in neurological disorders (Cryan et al. 2020), including Alzheimer's disease (e.g., Vogt et al. 2017; Liu et al. 2019).

As stated by Blaser and Falkow (2009), alterations in human macroecology have progressively affected the composition of our indigenous microbiota, affecting human physiology and disease risk. Therefore, these authors underline that the important factor in modern allergic and metabolic diseases might not be a decreased sampling of the microorganisms in food, air, water, or soil but instead could reflect the loss of our ancestral microorganisms. Münger et al. (2018), based on evidence emerging from biomedical and ecological studies, considered that the microbes, notably the gut microbiome and host social behaviour, coevolved to become virtually inseparable. The gut microbiome appears to shape their hosts' social behaviour and structure and depend on them for transmission. As summarised in Wilmes et al. (2022) review,

> The human gut microbiome produces a functional complex of biomolecules, including nucleic acids, (poly) peptides, structural molecules, and metabolites. This impacts human physiology in multiple ways... The gut microbiome must therefore be considered a central hub of exposures, which integrates environmental inputs with genetic and immune signals to affect host physiology.

Thus, studies in humans and laboratory animals suggest that the enteric microbial community plays a central role in brain function and cognitive development. The latter would result from the interaction of several domains, such as genetics, pregnancy and early developmental conditions, maternal and postnatal socioeconomic conditions, diet, and gut microbiome composition. Therefore, the complex interactions that ensue impact the context of the social brain hypothesis proposed by Dunbar (1998), affecting cognitive domains. Additional evidence demonstrates that microbes and probiotics can manipulate the social behaviour of their hosts, including altruism (Klein 2003; Bravo et al. 2011; Lewin-Epstein et al. 2017) and suggests that such manipulation has been subject to natural selection. As stated by Mayer et al. (2014),

> The initial skepticism about reports suggesting a profound role of an intact gut microbiota in shaping brain neurochemistry and emotional behaviour has given way to an unprecedented paradigm shift in the conceptualization of many psychiatric and neurological diseases. Although many of the new concepts are primarily based on the intriguing experimental findings in rodents, initial studies in humans seem to support the notion that there is a relationship between the complex world of microbiota in our intestines and brain structure and function.

As stated by Stilling et al. (2014) in a comprehensive review, selection favours social complexity to increase the transmission of beneficial microbes involved in host–symbiont coevolution impact on RNA regulation in the brain. This concept is reinforced by the observations of Pinacho-Guendulain et al. (2022) in non-human primates, suggesting a close connection between gut microbiota and social complexity within primate groups. The authors further state that social organisation, social structure, mating, and care systems may shape the gut microbiota composition at the group level.

On cognitive domains, according to Cooke et al. (2022), evidence suggests that gut microbiota is linked to brain connectivity and cognitive performance. Additionally, modulation of gut microbiota could be a promising strategy for enhancing cognition and emotional well-being in stressed and non-stressed situations. Following magnetic resonance imaging studies, the authors concluded that significant relationships exist between gut and oral microbiota diversity and resting state functional connectivity involving the insula, frontal pole, and left angular gyrus. These brain regions involve neural circuits integrating and processing information related to recognition, attention, and cognitive functions.

However, and most importantly, as commented by Mayer et al. (2014) and Morais et al. (2021) on behavioural grounds, since most of the studies that causally implicate gut microorganisms have used rodent models, further validation must be required before translation to human behaviour.

*

A Human Experiment in Spaceflight Conditions

Spaceflight conditions provide special settings to explore host-gut microbiome interactions in humans. Though several biomedical aspects emerge under such special living conditions, those related to the gut microbiome will be considered here as a special case of host-microbiome interactions. Astronauts encounter a range of conditions that may impact microbiome health, such as social isolation from the general population and extensive hygiene protocols that could impair microbiome diversity (Kuehnast et al. 2022). These authors also called for the need for spaceship structural integrity and planetary security from organisms known as extremophiles, i.e., those able to survive under extreme conditions. These biofilm-associated microorganisms exist in extreme environments and could degrade spacecraft materials. In addition to the initial screening, the spacecraft microbiome must be profiled before returning to Earth.

On this domain, Siddiqui et al. (2021) added,

> Overall, spaceflight has significant effects on the alteration of human microbiota and immune system of astronauts. Space travel changes the physiology as well as the composition of gut microbiome.

Brereton et al. (2021) reassessed common microbiome changes over time during the ground based Mars500 experiment following the isolation of six astronauts for 520 days. Findings revealed significantly altered microbiota in its relative abundance throughout confinement, including species known to affect inflammation and glucose homeostasis, consistent with astronaut symptoms. Voorhies et al. (2019) investigated the impact of long-term space exploration on the microbiome of nine astronauts who spent six to twelve months in the *International Space Station*. As stated by these authors, the gastrointestinal tract, skin, nose, and tongue microbial communities changed during the space mission. The intestinal microbiota composition became more similar across astronauts in space, primarily due to a drop in the abundance of a few bacterial taxa, some of which were also correlated with changes in the cytokine profile of crewmembers. In addition to physiological changes and reductions in immune response experienced by astronauts, microgravity can also cause changes in homeostasis and microbial biochemistry. The analysis identified 17 gastrointestinal genera whose abundance significantly changed in space, which returned to normality following the return to Earth, except for two genera of the phylum *Firmicutes* (many of them involved in producing butyrate, important for colon health). As stated by Turroni et al. (2020), studies available to date show that the space *exposome* –

the accumulated space travel-associated stressors – can strongly influence the gut microbiota with the potential impairment of the homeostatic relationship with the host. According to Tesei et al. (2022), spaceflight conditions promote alterations in the microbiome patterns (dysbiosis), determining shifts in the host microbiome interactions,

> Alterations in the composition and functionality of the gut microbiome can be induced even by short-term space travel. Liu et al. (2020) reported shifts between dominant genera in the microbiome during space missions of 15 and 35 days that led to increased abundance of Bacteroides. By contrast, the probiotic taxa *Lactobacillus* and *Bifidobacterium* appeared reduced, possibly affecting host immune function.

According to a NASA Report (2020)[1], most astronauts' gut microbiomes became more diverse in space rather than less, despite the limited exposure to new bacteria that may have resulted from the less variety of foods available on the station. It ought to be stressed that such changes in diversity involve both increases and reductions of different bacterial components.

Note

1 https://www.nasa.gov/humans-in-space/astronauts-leave-microbial-fingerprint-on-space-station/.

References

Allen, Andrew P., *et al.* "*Bifidobacterium Longum* 1714 as a Translational Psychobiotic: Modulation of Stress, Electrophysiology and Neurocognition in Healthy Volunteers." *Translational Psychiatry*, vol. 6, no. 11, 2016, doi:10.1038/tp.2016.191.

Allen, Andrew P., *et al.* "A Psychology of the Human Brain–Gut–Microbiome Axis." *Social and Personality Psychology Compass*, vol. 11, no. 4, 2017, doi:10.1111/spc3.12309.

Amato, Katherine R. "Incorporating the Gut Microbiota into Models of Human and Non-Human Primate Ecology and Evolution." *American Journal of Physical Anthropology*, vol. 159, no. S61, 2016, pp. 196–215, doi:10.1002/ajpa.22908.

Angoorani, Pooneh, *et al.* "Is there any Link between Cognitive Impairment and Gut Microbiota? A Systematic Review." *Gerontology*, vol. 68, no. 11, 2022, pp. 1201–1213, doi:10.1159/000522381.

Archie, Elizabeth A., and Jenny Tung. "Social Behavior and the Microbiome." *Current Opinion in Behavioral Sciences*, vol. 6, 2015, pp. 28–34, doi:10.1016/j.cobeha.2015.07.008.

Arentsen, Tim, *et al.* "Host Microbiota Modulates Development of Social Preference in Mice." *Microbial Ecology in Health & Disease*, vol. 26, 2015.

Arumugam, Manimozhiyan, *et al.* "Enterotypes of the Human Gut Microbiome." *Nature*, vol. 473, no. 7346, 2011, pp. 174–180, doi:10.1038/nature09944.

Bailey, Michael T., *et al.* "Exposure to a Social Stressor Alters the Structure of the Intestinal Microbiota: Implications for Stressor-Induced Immunomodulation." *Brain, Behavior, and Immunity*, vol. 25, no. 3, 2011, pp. 397–407, doi:10.1016/j.bbi.2010.10.023.

Baldi, Simone, *et al.* "Microbiota Shaping – the Effects of Probiotics, Prebiotics, and Fecal Microbiota Transplant on Cognitive Functions: A Systematic Review." *World Journal of Gastroenterology*, vol. 27, no. 39, 2021, pp. 6715–6732, doi:10.3748/wjg.v27.i39.6715.

Bauer, Kylynda C., *et al.* "The Gut Microbiota–Brain Axis Expands Neurologic Function: A Nervous Rapport." *BioEssays*, vol. 41, no. 10, 2019, doi:10.1002/bies.201800268.

Bercik, Premysl, *et al.* "The Intestinal Microbiota Affect Central Levels of Brain-Derived Neurotropic Factor and Behavior in Mice." *Gastroenterology*, vol. 141, no. 2, 2011, doi:10.1053/j.gastro.2011.04.052.

Bercik, Premysl, *et al.* "Microbes and the Gut-Brain Axis." *Neurogastroenterology & Motility*, vol. 24, no. 5, 2012, pp. 405–413, doi:10.1111/j.1365-2982.2012.01906.x.

Blaser, Martin J., and Stanley Falkow. "What are the Consequences of the Disappearing Human Microbiota?" *Nature Reviews Microbiology*, vol. 7, no. 12, 2009, pp. 887–894, doi:10.1038/nrmicro2245.

Borre, Yuliya E., *et al.* "The Impact of Microbiota on Brain and Behavior: Mechanisms & Therapeutic Potential." In *Microbial Endocrinology: The Microbiota-Gut-Brain Axis in Health and Disease*, edited by Mark Lyte and John F. Cryan. New York: Springer, 2014, pp. 373–403, doi:10.1007/978-1-4939-0897-4_17.

Bravo, Javier A., *et al.* "Ingestion of *Lactobacillus* Strain Regulates Emotional Behavior and Central GABA Receptor Expression in a Mouse via the Vagus Nerve." *Proceedings of the National Academy of Sciences*, vol. 108, no. 38, 2011, pp. 16050–16055, doi:10.1073/pnas.1102999108.

Brereton, N. J. B., *et al.* "Reanalysis of the MARS500 Experiment Reveals Common Gut Microbiome Alterations in Astronauts Induced by Long-Duration Confinement." *Computational and Structural Biotechnology Journal*, vol. 19, 2021, pp. 2223–2235, doi:10.1016/j.csbj.2021.03.040.

Brown, Eric M., *et al.* "Gut Microbiome ADP-Ribosyltransferases are Widespread Phage-Encoded Fitness Factors." *Cell Host & Microbe*, vol. 29, no. 9, 2021, pp. 1351–1365, doi:10.1016/j.chom.2021.07.011.

Bruckner, Joseph J., *et al.* "The Microbiota Promotes Social Behavior by Modulating Microglial Remodeling of Forebrain Neurons." *PLoS Biology*, vol. 20, no. 11, 2022, doi:10.1371/journal.pbio.3001838.

Buffington, Shelly A., *et al.* "Microbial Reconstitution Reverses Maternal Diet-Induced Social and Synaptic Deficits in Offspring." *Cell*, vol. 165, no. 7, 2016, pp. 1762–1775, doi:10.1016/j.cell.2016.06.001.

Cani, Patrice D., and Claude Knauf. "How Gut Microbes Talk to Organs: The Role of Endocrine and Nervous Routes." *Molecular Metabolism*, vol. 5, no. 9, 2016, pp. 743–752, doi:10.1016/j.molmet.2016.05.011.

Carabotti, Marilia, *et al.* "The Gut-Brain Axis: Interactions between Enteric Microbiota, Central and Enteric Nervous Systems." *Annals of Gastroenterology*, vol. 28, no. 2, 2015, pp. 203–209.

Carlson, Alexander L., *et al.* "Infant Gut Microbiome Associated with Cognitive Development." *Biological Psychiatry*, vol. 83, no. 2, 2018, pp. 148–159, doi:10.1016/j.biopsych.2017.06.021.

Carrier, Tyler J., and Adam M. Reitzel. "The Hologenome Across Environments and the Implications of a Host-Associated Microbial Repertoire." *Frontiers in Microbiology*, vol. 8, 2017, doi:10.3389/fmicb.2017.00802.

Chang, Chia-Yu, *et al.* "Essential Fatty Acids and Human Brain." *Acta neurologica Taiwanica*, vol. 18, no. 4, 2009, pp. 231–241.

Chen, Zilin, *et al.* "Gut Microbial Profile Is Associated with the Severity of Social Impairment and IQ Performance in Children with Autism Spectrum Disorder." *Frontiers in Psychiatry*, vol. 12, 2021, doi:10.3389/fpsyt.2021.789864.

Clarke, Gerard, and John F. Cryan. "Preface." In *International Review of Neurobiology: Gut Microbiome and Behavior, Vol. 131*, edited by John F. Cryan and Gerard Clarke, Cambridge, MA: Elsevier, 2016, pp. xv–xxiii.

Collins, Stephen M., *et al.* "The Adoptive Transfer of Behavioral Phenotype via the Intestinal Microbiota: Experimental Evidence and Clinical Implications." *Current Opinion in Microbiology*, vol. 16, no. 3, 2013, pp. 240–245, doi:10.1016/j.mib.2013.06.004.

Cani, Patrice D., and Claude Knauf. "How Gut Microbes Talk to Organs: The Role of Endocrine and Nervous Routes." *Molecular Metabolism*, vol. 5, no. 9, 2016, pp. 743–752, doi:10.1016/j.molmet.2016.05.011.

Cooke, Matthew B., *et al.* "Examining the Influence of the Human Gut Microbiota on Cognition and Stress: A Systematic Review of the Literature." *Nutrients*, vol. 14, no. 21, 2022, p. 4623, doi:10.3390/nu14214623.

Cryan, John F., and Timothy G. Dinan. "Mind-Altering Microorganisms: The Impact of the Gut Microbiota on Brain and Behaviour." *Nature Reviews Neuroscience*, vol. 13, no. 10, 2012, pp. 701–712, www.nature.com/articles/nrn3346, doi:10.1038/nrn3346.

Cryan, John F, *et al.* "The Gut Microbiome in Neurological Disorders." *The Lancet Neurology*, vol. 19, no. 2, 2020, www.sciencedirect.com/science/article/pii/S1474442219303564, doi:10.1016/s1474-4422(19)30356-4.

D'Amato, Alfonsina, *et al.* "Faecal Microbiota Transplant from Aged Donor Mice Affects Spatial Learning and Memory via Modulating Hippocampal Synaptic Plasticity- and Neurotransmission-Related Proteins in Young Recipients." *Microbiome*, vol. 8, no. 140, 2020, doi:10.1186/s40168-020-00914-w.

De Filippo, C., *et al.* "Impact of Diet in Shaping Gut Microbiota Revealed by a Comparative Study in Children from Europe and Rural Africa." *Proceedings of the National Academy of Sciences*, vol. 107, no. 33, 2010, pp. 14691–14696, www.pnas.org/content/107/33/14691.full, doi:10.1073/pnas.1005963107.

Degnan, P. H., *et al.* "Factors Associated with the Diversification of the Gut Microbial Communities within Chimpanzees from Gombe National Park." *Proceedings of the National Academy of Sciences*, vol. 109, no. 32, 2012, pp. 13034–13039, doi:10.1073/pnas.1110994109.

Desbonnet, Lieve, *et al.* "Microbiota Is Essential for Social Development in the Mouse." *Molecular Psychiatry*, vol. 19, no. 2, 2014, pp. 146–148, doi:10.1038/mp.2013.65.

Desbonnet, Lieve, *et al.* "Gut Microbiota Depletion from Early Adolescence in Mice: Implications for Brain and Behaviour." *Brain, Behavior, and Immunity*, vol. 48, 2015, pp. 165–173, doi:10.1016/j.bbi.2015.04.004.

De Vadder, Filipe, *et al.* "Microbiota-Generated Metabolites Promote Metabolic Benefits via Gut-Brain Neural Circuits." *Cell*, vol. 156, no. 1–2, 2014, pp. 84–96, www.sciencedirect.com/science/article/pii/S009286741301550X, doi:10.1016/j.cell.2013.12.016.

Dinan, Timothy G., *et al.* "Collective Unconscious: How Gut Microbes Shape Human Behavior." *Journal of Psychiatric Research*, vol. 63, 2015, pp. 1–9, doi:10.1016/j.jpsychires.2015.02.021.

Dunbar, Robin I. M. "The Social Brain Hypothesis." *Evolutionary Anthropology: Issues, News, and Reviews*, vol. 6, no. 5, 1998, pp. 178–190, doi:10.1002/(SICI)1520-6505(1998)6:5%3C178::AID-EVAN5%3E3.0.CO;2-8.

Eloe-Fadrosh, Emiley A., and David A. Rasko. "The Human Microbiome: From Symbiosis to Pathogenesis." *Annual Review of Medicine*, vol. 64, no. 1, 2013, pp. 145–163, www.ncbi.nlm.nih.gov/pmc/articles/PMC3731629/, doi:10.1146/annurev-med-010312-133513.

Ezenwa, V. O., et al. "Animal Behavior and the Microbiome." *Science*, vol. 338, no. 6104, 11 Oct. 2012, pp. 198–199, doi:10.1126/science.1227412.

Felice, Valeria D., and Siobhain M. O'Mahony. "The Microbiome and Disorders of the Central Nervous System." *Pharmacology Biochemistry and Behavior*, vol. 160, 2017, pp. 1–13, doi:10.1016/j.pbb.2017.06.016.

Forsythe, Paul, *et al.* "Mood and Gut Feelings." *Brain, Behavior, and Immunity*, vol. 24, no. 1, 2010, pp. 9–16, doi:10.1016/j.bbi.2009.05.058.

Foster, Jane A., and Karen-Anne McVey Neufeld. "Gut–Brain Axis: How the Microbiome Influences Anxiety and Depression." *Trends in Neurosciences*, vol. 36, no. 5, 2013, pp. 305–312.

Foster, Jane A, *et al.* "Gut Microbiota and Brain Function: An Evolving Field in Neuroscience: Table 1." *International Journal of Neuropsychopharmacology*, vol. 19, no. 5, 2015, p. pyv114, academic.oup.com/ijnp/article/19/5/pyv114/2910042, doi:10.1093/ijnp/pyv114.

Frank, Daniel N., and Norman R. Pace. "Gastrointestinal Microbiology Enters the Metagenomics Era." *Current Opinion in Gastroenterology*, vol. 24, no. 1, 2008, pp. 4–10, doi:10.1097/mog.0b013e3282f2b0e8.

Furness, John B. "The Enteric Nervous System and Neurogastroenterology." *Nature Reviews Gastroenterology & Hepatology*, vol. 9, no. 5, 2012, pp. 286–294, doi:10.1038/nrgastro.2012.32.

Gareau, M. G., et al. "Bacterial Infection Causes Stress-Induced Memory Dysfunction in Mice." *Gut*, vol. 60, no. 3, 2011, pp. 307–317, doi:10.1136/gut.2009.202515.

Gareau, M. G. "Cognitive Function and the Microbiome." In *International Review of Neurobiology: Gut Microbiome and Behavior, Vol 131*, edited by John F. Cryan and Gerard Clarke, Cambridge, MA: Elsevier, 2016, pp. 227–246, doi:10.1016/bs.irn.2016.08.001.

Gacias, Mar, et al. "Microbiota-Driven Transcriptional Changes in Prefrontal Cortex Override Genetic Differences in Social Behavior." *ELife*, vol. 5, 2016, doi:10.7554/elife.13442.

Garud, Nandita R., et al. "Evolutionary Dynamics of Bacteria in the Gut Microbiome within and across Hosts." *PLoS Biology*, vol. 17, no. 1, 2019, p. e3000102, doi:10.1371/journal.pbio.3000102.

Goyal, Manu, et al. "Feeding the Brain and Nurturing the Mind: Linking Nutrition and the Gut Microbiota to Brain Development." *Proceedings of the National Academy of Sciences of the United States of America*, vol. 112, no. 46, 2015, pp. 14105–14112, doi:10.1073/pnas.1511465112.

Hanstock, T., et al. "Anxiety and Aggression Associated with the Fermentation of Carbohydrates in the Hindgut of Rats." *Physiology & Behavior*, vol. 82, no. 2–3, 2004, pp. 357–368, doi:10.1016/j.physbeh.2004.04.002.

Heijtz, R. D., et al. "Normal Gut Microbiota Modulates Brain Development and Behavior." *Proceedings of the National Academy of Sciences*, vol. 108, no. 7, 2011, pp. 3047–3052, doi:10.1073/pnas.1010529108.

Hooper, Lora V. et al. "How Host-Microbial Interactions Shape the Nutrient Environment of the Mammalian Intestine." *Annual Review of Nutrition*, vol. 22, 2002, pp. 283–307. doi:10.1146/annurev.nutr.22.011602.092259.

Hooper, Lora V. "Bacterial Contributions to Mammalian Gut Development." *Trends in Microbiology*, vol. 12, no. 3, 2004, pp. 129–134, www.sciencedirect.com/science/article/pii/S0966842X04000150, doi:10.1016/j.tim.2004.01.001.

Hsiao, Elaine Y. et al. "Microbiota Modulate Behavioral and Physiological Abnormalities Associated with Neurodevelopmental Disorders." *Cell*, vol. 155, no. 7, 2013, pp. 1451–1463, doi:10.1016/j.cell.2013.11.024.

Iyer, Lakshminarayan M., et al. "Evolution of Cell–Cell Signaling in Animals: Did Late Horizontal Gene Transfer from Bacteria Have a Role?" *Trends in Genetics*, vol. 20, no. 7, 2004, pp. 292–299, doi:10.1016/j.tig.2004.05.007.

Jašarević, Eldin, et al. "Sex Differences in the Gut Microbiome–Brain Axis across the Lifespan." *Philosophical Transactions of the Royal Society B: Biological Sciences*, vol. 371, no. 1688, 2016, p. 20150122, doi:10.1098/rstb.2015.0122.

Johnson, Katerina V. A., and Kevin R. Foster. "Why Does the Microbiome Affect Behaviour?" *Nature Reviews Microbiology*, vol. 16, no. 10, 2018, pp. 647–655, doi:10.1038/s41579-018-0014-3.

Johnson, Katerina V. A. "Gut Microbiome Composition and Diversity Are Related to Human Personality Traits." *Human Microbiome Journal*, vol. 15, 2020, p. 100069, www.sciencedirect.com/science/article/pii/S2452231719300181, doi:10.1016/j.humic.2019.100069.

Kamiya, T., et al. "Inhibitory Effects of *Lactobacillus Reuteri* on Visceral Pain Induced by Colorectal Distension in Sprague-Dawley Rats." *Gut*, vol. 55, no. 2, 2006, pp. 191–196, doi:10.1136/gut.2005.070987.

Klein, Sabra L. "Parasite Manipulation of the Proximate Mechanisms That Mediate Social Behavior in Vertebrates." *Physiology & Behavior*, vol. 79, no. 3, 2003, pp. 441–449, doi:10.1016/s0031-9384(03)00163-x.

Kuehnast, Torben, et al. "The Crewed Journey to Mars and Its Implications for the Human Microbiome." *Microbiome*, vol. 10, no. 26, 2022, doi:10.1186/s40168-021-01222-7.

Kurokawa, Ken, *et al.* "Comparative Metagenomics Revealed Commonly Enriched Gene Sets in Human Gut Microbiomes." *DNA Research*, vol. 14, no. 4, 2007, pp. 169–181, doi:10.1093/dnares/dsm018.

Kurokawa, Shunya, *et al.* "The Effect of Fecal Microbiota Transplantation on Psychiatric Symptoms among Patients with Irritable Bowel Syndrome, Functional Diarrhea and Functional Constipation: An Open-Label Observational Study." *Journal of Affective Disorders*, vol. 235, 2018, pp. 506–512, pubmed.ncbi.nlm.nih.gov/29684865/, doi:10.1016/j.jad.2018.04.038.

Lewin-Epstein, Ohad, *et al.* "Microbes Can Help Explain the Evolution of Host Altruism." *Nature Communications*, vol. 8, no. 1, 2017, doi:10.1038/ncomms14040.

Li, Wang, *et al.* "Memory and Learning Behavior in Mice Is Temporally Associated with Diet-Induced Alterations in Gut Bacteria." *Physiology & Behavior*, vol. 96, no. 4–5, 2009, pp. 557–567, doi:10.1016/j.physbeh.2008.12.004.

Liang, Shan, *et al.* "Editorial: The Effect of Gut Microbiota on the Brain Structure and Function." *Frontiers in Integrative Neuroscience*, vol. 17, 2023, doi:10.3389/fnint.2023.1226664.

Liang, Xinxiu, *et al.* "Gut Microbiome, Cognitive Function and Brain Structure: A Multi-Omics Integration Analysis." *Translational Neurodegeneration*, vol. 11, no. 49, 2022, www.ncbi.nlm.nih.gov/pmc/articles/PMC9661756/, doi:10.1186/s40035-022-00323-z.

Liu, Ping, *et al.* "Altered Microbiomes Distinguish Alzheimer's Disease from Amnestic Mild Cognitive Impairment and Health in a Chinese Cohort." *Brain, Behavior, and Immunity*, vol. 80, 2019, pp. 633–643, doi:10.1016/j.bbi.2019.05.008.

Lombardo, Michael P. "Access to Mutualistic Endosymbiotic Microbes: An Underappreciated Benefit of Group Living." *Behavioral Ecology and Sociobiology*, vol. 62, no. 4, 2007, pp. 479–497, doi:10.1007/s00265-007-0428-9.

Louwies, Tijs, *et al.* "The Microbiota-Gut-Brain Axis: An Emerging Role for the Epigenome." *Experimental Biology and Medicine*, vol. 245, no. 2, 2020, pp. 138–145, doi:10.1177/1535370219891690.

Lu, Jing, *et al.* "Effects of Intestinal Microbiota on Brain Development in Humanized Gnotobiotic Mice." *Scientific Reports*, vol. 8, no. 1, 2018, doi:10.1038/s41598-018-23692-w.

Luczynski, Pauline, *et al.* "Adult Microbiota-Deficient Mice Have Distinct Dendritic Morphological Changes: Differential Effects in the Amygdala and Hippocampus." *European Journal of Neuroscience*, vol. 44, no. 9, 2016, pp. 2654–2666, doi:10.1111/ejn.13291.

Lyte, Mark. "The Role of Microbial Endocrinology in Infectious Disease." *Journal of Endocrinology*, vol. 137, no. 3, 1993, pp. 343–345, doi:10.1677/joe.0.1370343.

Lyte, Mark. "The Microbial Organ in the Gut as a Driver of Homeostasis and Disease." *Medical Hypotheses*, vol. 74, no. 4, 2010, pp. 634–638, doi:10.1016/j.mehy.2009.10.025.

Lyte, Mark. "Microbial Endocrinology in the Microbiome-Gut-Brain Axis: How Bacterial Production and Utilization of Neurochemicals Influence Behavior." *PLoS Pathogens*, vol. 9, no. 11, 2013, p. e1003726, doi:10.1371/journal.ppat.1003726.

Mayer, Emeran A. "Gut Feelings: The Emerging Biology of Gut–Brain Communication." *Nature Reviews Neuroscience*, vol. 12, no. 8, 2011, pp. 453–466, www.nature.com/articles/nrn3071, doi:10.1038/nrn3071.

Mayer, Emeran A., *et al.* "Gut Microbes and the Brain: Paradigm Shift in Neuroscience." *Journal of Neuroscience*, vol. 34, no. 46, 2014, pp. 15490–15496, www.jneurosci.org/content/34/46/15490.short, doi:10.1523/jneurosci.3299-14.2014.

McGuinness, A. J., *et al.* "A Systematic Review of Gut Microbiota Composition in Observational Studies of Major Depressive Disorder, Bipolar Disorder and Schizophrenia." *Molecular Psychiatry*, vol. 27, no. 4, 2022, pp. 1920–1935, www.nature.com/articles/s41380-022-01456-3, doi:10.1038/s41380-022-01456-3.

McKernan, D. P., *et al.* "The Probiotic *Bifidobacterium Infantis* 35624 Displays Visceral Antinociceptive Effects in the Rat." *Neurogastroenterology & Motility*, vol. 22, no. 9, 2010, pp. 1029–e268, doi:10.1111/j.1365-2982.2010.01520.x.

Messaoudi, Michaël, *et al.* "Beneficial Psychological Effects of a Probiotic Formulation (*Lactobacillus Helveticus* R0052 and *Bifidobacterium Longum* R0175) in Healthy Human Volunteers." *Gut Microbes*, vol. 2, no. 4, 2011, pp. 256–261, doi:10.4161/gmic.2.4.16108.

Miller, Gregory E., *et al.* "Lower Neighborhood Socioeconomic Status Associated with Reduced Diversity of the Colonic Microbiota in Healthy Adults." *PLoS ONE*, vol. 11, no. 2, 2016, p. e0148952, doi:10.1371/journal.pone.0148952.

Moeller, Andrew H., *et al.* "Social Behavior Shapes the Chimpanzee Pan-Microbiome." *Science Advances*, vol. 2, no. 1, 2016, p. e1500997, advances.sciencemag.org/content/2/1/e1500997, doi:10.1126/sciadv.1500997.

Mohajeri, M. Hasan, *et al.* "Relationship between the Gut Microbiome and Brain Function." *Nutrition Reviews*, vol. 76, no. 7, 2018, pp. 481–496, doi:10.1093/nutrit/nuy009.

Montiel-Castro, Augusto J.*et al.* "The Microbiota-Gut-Brain Axis: Neurobehavioral Correlates, Health and Sociality." *Frontiers in Integrative Neuroscience*, vol. 7, 2013, doi:10.3389/fnint.2013.00070.

Morais, Livia H.*et al.* "The Gut Microbiota-Brain Axis in Behaviour and Brain Disorders." *Nature Reviews Microbiology*, vol. 19, no. 4, 2021, pp. 241–255, doi:10.1038/s41579-020-00460-0.

Münger, Emmanuelle, *et al.* "Reciprocal Interactions between Gut Microbiota and Host Social Behavior." *Frontiers in Integrative Neuroscience*, vol. 12, 2018, doi:10.3389/fnint.2018.00021.

NASA, "Astronauts Leave 'Microbial Fingerprint' on Space Station." *NASA Report*, 28 July 2023, www.nasa.gov/feature/ames/astronauts-leave-microbial-fingerprint-on-space-station/.

Neufeld, Karen-Anne M., *et al.* "Effects of Intestinal Microbiota on Anxiety-Like Behavior." *Communicative & Integrative Biology*, vol. 4, no. 4, 2011, pp. 492–494. doi:10.4161/cib.4.4.15702.

Nikolova, Viktoriya L., *et al.* "Perturbations in Gut Microbiota Composition in Psychiatric Disorders." *JAMA Psychiatry*, vol. 78, no. 12, 2021, pp. 1343–1354, doi:10.1001/jamapsychiatry.2021.2573.

Ochman, Howard, *et al.* "Evolutionary Relationships of Wild Hominids Recapitulated by Gut Microbial Communities." *PLoS Biology*, vol. 8, no. 11, 2010, p. e1000546, doi:10.1371/journal.pbio.1000546.

O'Mahony, S. M., *et al.* "Serotonin, Tryptophan Metabolism and the Brain-Gut-Microbiome Axis." *Behavioural Brain Research*, vol. 277, 2015, pp. 32–48, doi:10.1016/j.bbr.2014.07.027.

Oroojzadeh, Parvin, *et al.* "Psychobiotics: The Influence of Gut Microbiota on the Gut-Brain Axis in Neurological Disorders." *Journal of Molecular Neuroscience*, vol. 72, 2022, pp. 1952–1964, doi:10.1007/s12031-022-02053-3.

Palepu, M. S. K., and M. P. Dandekar. "Remodeling of Microbiota Gut-Brain Axis using Psychobiotics in Depression." *European Journal of Pharmacology*, vol. 931, no. 175171, 2022, doi:10.1016/j.ejphar.2022.175171.

Parsons, Emilee, *et al.* "The Infant Microbiome and Implications for Central Nervous System Development." *Progress in Molecular Biology and Translational Science*, vol. 171, 2020, pp. 1–13, pubmed.ncbi.nlm.nih.gov/32475519, doi:10.1016/bs.pmbts.2020.04.007.

Pasquaretta, Cristian, *et al.* "Exploring Interactions between the Gut Microbiota and Social Behavior through Nutrition." *Genes*, vol. 9, no. 11, 2018, p. 534, doi:10.3390/genes9110534.

Pinacho-Guendulain, Braulio, *et al.* "Social Complexity as a Driving Force of Gut Microbiota Exchange among Conspecific Hosts in Non-Human Primates." *Frontiers in Integrative Neuroscience*, vol. 16, 2022, doi:10.3389/fnint.2022.876849.

Safadi, Jenelle Marcelle, *et al.* "Gut Dysbiosis in Severe Mental Illness and Chronic Fatigue: A Novel Trans-Diagnostic Construct? A Systematic Review and Meta-Analysis." *Molecular Psychiatry*, vol. 27, 2022, pp. 141–153, www.nature.com/articles/s41380-021-01032-1, doi:10.1038/s41380-021-01032-1.

Sampson, Timothy R., and Sarkis K. Mazmanian. "Control of Brain Development, Function, and Behavior by the Microbiome." *Cell Host & Microbe*, vol. 17, no. 5, 2015, pp. 565–576, doi:10.1016/j.chom.2015.04.011.

Sarkar, Amar, *et al.* "The Role of the Microbiome in the Neurobiology of Social Behaviour." *Biological Reviews*, vol. 95, no. 5, 2020, pp. 1131–1166, doi:10.1111/brv.12603.

Savage, D. C. "Gastrointestinal Microflora in Mammalian Nutrition." *Annual Review of Nutrition*, vol. 6, no. 1, 1986, pp. 155–178, doi:10.1146/annurev.nu.06.070186.001103.

Sender, Ron, *et al.* "Revised Estimates for the Number of Human and Bacteria Cells in the Body." *PLoS Biology*, vol. 14, no. 8, 2016, p. e1002533, doi:10.1371/journal.pbio.1002533.

Sharon, Gil, *et al.* "The Central Nervous System and the Gut Microbiome." *Cell*, vol. 167, no. 4, 2016, pp. 915–932, doi:10.1016/j.cell.2016.10.027.

Shoubridge, Andrew P., *et al.* "The Gut Microbiome and Mental Health: Advances in Research and Emerging Priorities." *Molecular Psychiatry*, vol. 27, 2022, pp. 1908–1919, doi:10.1038/s41380-022-01479-w.

Shroff, K. E., *et al.* "Commensal Enteric Bacteria Engender a Self-Limiting Humoral Mucosal Immune Response While Permanently Colonizing the Gut." *Infection and Immunity*, vol. 63, no. 10, 1995, pp. 3904–3913, doi:10.1128/iai.63.10.3904-3913.1995.

Siddiqui, R., *et al.* "Gut Microbiome and Human Health under the Space Environment." *Journal of Applied Microbiology*, vol. 130, no. 1, 2021, pp. 14–24, doi:10.1111/jam.14789.

Stilling, Roman M., *et al.* "Friends with Social Benefits: Host-Microbe Interactions as a Driver of Brain Evolution and Development?" *Frontiers in Cellular and Infection Microbiology*, vol. 4, no. 147, 2014, doi:10.3389/fcimb.2014.00147.

Stilling, Roman M., *et al.* "The Brain's Geppetto—Microbes as Puppeteers of Neural Function and Behaviour?" *Journal of NeuroVirology*, vol. 22, no. 1, 2015, pp. 14–21, doi:10.1007/s13365-015-0355-x.

Sudo, Nobuyuki, *et al.* "Postnatal Microbial Colonization Programs the Hypothalamic-Pituitary-Adrenal System for Stress Response in Mice." *The Journal of Physiology*, vol. 558, no. 1, 2004, pp. 263–275, doi:10.1113/jphysiol.2004.063388.

Suganya, Kanmani, and Byung-Soo Koo. "Gut–Brain Axis: Role of Gut Microbiota on Neurological Disorders and How Probiotics/Prebiotics Beneficially Modulate Microbial and Immune Pathways to Improve Brain Functions." *International Journal of Molecular Sciences*, vol. 21, no. 20, 2020, p. 7551, www.mdpi.com/1422-0067/21/20/7551/htm, doi:10.3390/ijms21207551.

Tengeler, Anouk C., *et al.* "Gut Microbiota from Persons with Attention-Deficit/Hyperactivity Disorder Affects the Brain in Mice." *Microbiome*, vol. 8, no. 44, 2020, doi:10.1186/s40168-020-00816-x.

Tesei, Donatella, *et al.* "Understanding the Complexities and Changes of the Astronaut Microbiome for Successful Long-Duration Space Missions." *Life*, vol. 12, no. 4, 2022, p. 495, doi:10.3390/life12040495.

Tillisch, Kirsten, *et al.* "Consumption of Fermented Milk Product with Probiotic Modulates Brain Activity." *Gastroenterology*, vol. 144, no. 7, 2013, pp. 1394–1401.e4, www.ncbi.nlm.nih.gov/pmc/articles/PMC3839572/, doi:10.1053/j.gastro.2013.02.043.

Tung, Jenny, *et al.* "Social Networks Predict Gut Microbiome Composition in Wild Baboons." *ELife*, vol. 4, 2015, doi:10.7554/elife.05224.

Turroni, Silvia, *et al.* "Gut Microbiome and Space Travelers' Health: State of the Art and Possible Pro/Prebiotic Strategies for Long-Term Space Missions." *Frontiers in Physiology*, vol. 11, 2020, doi:10.3389/fphys.2020.553929.

Valles-Colomer, Mireia, *et al.* "The Neuroactive Potential of the Human Gut Microbiota in Quality of Life and Depression." *Nature Microbiology*, vol. 4, no. 4, 2019, pp. 623–632, www.nature.com/articles/s41564-018-0337-x, doi:10.1038/s41564-018-0337-x.

Verdu, E. F. "Specific Probiotic Therapy Attenuates Antibiotic Induced Visceral Hypersensitivity in Mice." *Gut*, vol. 55, no. 2, 2006, pp. 182–190, doi:10.1136/gut.2005.066100.

Vogt, Nicholas M., *et al.* "Gut Microbiome Alterations in Alzheimer's Disease." *Scientific Reports*, vol. 7, no. 1, 2017, www.nature.com/articles/s41598-017-13601-y, doi:10.1038/s41598-017-13601-y.

Voorhies, Alexander A., *et al.* "Study of the Impact of Long-Duration Space Missions at the International Space Station on the Astronaut Microbiome." *Scientific Reports*, vol. 9, no. 1, 2019, www.nature.com/articles/s41598-019-46303-8, doi:10.1038/s41598-019-46303-8.

Vuong, Helen E., *et al.* "The Microbiome and Host Behavior." *Annual Review of Neuroscience*, vol. 40, no. 1, 2017, pp. 21–49, doi:10.1146/annurev-neuro-072116-031347.

Walter, Jens, and Ruth Ley. "The Human Gut Microbiome: Ecology and Recent Evolutionary Changes." *Annual Review of Microbiology*, vol. 65, no. 1, 2011, pp. 411–429, doi:10.1146/annurev-micro-090110-102830.

Wang, Yan, and Lloyd H. Kasper. "The Role of Microbiome in Central Nervous System Disorders." *Brain, Behavior, and Immunity*, vol. 38, 2014, pp. 1–12, doi:10.1016/j.bbi.2013.12.015.

Wilmes, Paul, *et al.* "The Gut Microbiome Molecular Complex in Human Health and Disease." *Cell Host & Microbe*, vol. 30, no. 9, 2022, pp. 1201–1206, doi:10.1016/j.chom.2022.08.016.

Yano, Jessica M., *et al.* "Indigenous Bacteria from the Gut Microbiota Regulate Host Serotonin Biosynthesis." *Cell*, vol. 161, no. 2, 2015, pp. 264–276, www.ncbi.nlm.nih.gov/pubmed/25860609, doi:10.1016/j.cell.2015.02.047.

Zilber-Rosenberg, Ilana, and Eugene Rosenberg. "Role of Microorganisms in the Evolution of Animals and Plants: The Hologenome Theory of Evolution." *FEMS Microbiology Reviews*, vol. 32, no. 5, 2008, pp. 723–735, doi:10.1111/j.1574-6976.2008.00123.x.

Zhu, Sibo, *et al.* "The Progress of Gut Microbiome Research Related to Brain Disorders." *Journal of Neuroinflammation*, vol. 17, no. 1, 2020, doi:10.1186/s12974-020-1705-z.

Zhu, Yao, *et al.* "Interactions between Intestinal Microbiota and Neural Mitochondria: A New Perspective on Communicating Pathway from Gut to Brain." *Frontiers in Microbiology*, vol. 13, 2022a, p. 798917, www.ncbi.nlm.nih.gov/pmc/articles/PMC8908256/, doi:10.3389/fmicb.2022.798917.

Zhu, Fangyuan, *et al.* "The Microbiota–Gut–Brain Axis in Depression: The Potential Pathophysiological Mechanisms and Microbiota Combined Antidepression Effect." *Nutrients*, vol. 14, no. 10, 2022b, p. 2081, doi:10.3390/nu14102081.

9

HUMAN POSTNATAL GROWTH AND *CRITICAL PERIOD*

Conditions of Human Postnatal Growth

Millions of years of evolution, of trial and error – of transient survival successes and failures – are expressed in each of us. We carry structures and functions needed under different environmental or living conditions, now modified or adapted to modern living, if not under limited, constrained conditions. Our species' presence on the planet is a process open to variables promoting either change or extinction and replacement, a notion that is antithetical to that which holds that humans represent the culmination of a supposed evolutionary pyramid or that are the product of an intelligent design.

Within the Animal Kingdom, securing enough territory for feeding and reproduction represents a priority, resulting in intraspecific belligerence due to competition. In gregarious communities with hierarchical organisation, access to nutrient resources is based on a hierarchical arrangement of rights. Hence, in this type of community, the hierarchical structure decides the possible levels of survival and welfare of its constituents, depending among other variables, on food supplies (Kerhoas et al. 2014). This social structure could result in competition for food rights and potentially terminal confrontations. Historically, food access to valuable nutrients has become a potential source of conflict. In our *sapiens* culture, among other strategic issues, access to drinking water and fertile lands tends to promote hostility or deadly wars. The advent of farming settlements and the exploitation control of land yield became a source of economic and political power and a private target that altered relative powers and confrontations profiles. In this regard, Kohler et al. (2017) discussed the disparity development of richness since the end of the *Neolithic* period in Eurasia, associated with land property and increased farming and domestication of livestock. The emergence of the economic concept of private property contributed with a new element to the construction of hierarchical societies and social inequalities.

Humans have a very extended period of postnatal helplessness compared to other species, during which our physical and mental development is completed under

DOI: 10.4324/9781032698380-9

conditions spanning through environmental cues. For this reason, the necessary breeding period imposed particularly demanding conditions on the emergence of *Homo sapiens*. It also required meeting the demands of a developing brain, which would continue its development during postnatal life.

The human brain is comparatively disproportionate compared to other mammals, considering weight and size. The level of cerebral metabolic demand is characteristic of the genus *Homo*. It is primarily due to our comparatively large cerebral cortex, which required folding in on itself inside a relatively inextensible skull during late prenatal life and early childhood. A larger head volume would have compromised even further child delivery. Undoubtedly, the human newborn posed a combination of variables with a higher level of physical and nutritional demand for the social group, such as extended caring for a child in a hostile environment. The risks such a combination would have posed for a species that typically produces only one offspring per pregnancy would have been very high.

The large development of the cerebral cortex has been related to the complex degree of socialisation of *Homo sapiens* and involved genetic factors (e.g., Thompson et al. 2001; Mekel-Bobrov et al. 2005; Evans et al. 2005). Thus, on the evolutionary horizon, the limits imposed by both dimensions – brain development and social complexities – carry much of the burden implied by the emergence of a neonatal *Homo sapiens*.

> Social status is one of the most important predictors of the quality of an individual's social environment.
>
> *(Tung et al. 2012)*

From an ethical and social point of view, one of the most pressing problems lies when childcare evolves under deficient conditions as an undesirable product due to inequality and social marginalisation. In these cases, the conditions of child upbringing impose negative biases on large sectors of the population on a series of fundamental aspects, including food supply that impinge on essential nutrients and gut microbiome development. These biases can perpetuate for a long time in degraded communities due to the lack of adequate public policies.

*

Birth Weight and Energy Reserves

Within an evolutive context, conditions of prolonged cerebral and neuromuscular immaturity of the human foetus, in all probability, would have implied a high protective demand against predators that would provide the prolonged postnatal development to which our species was subjected. Given the degree of defencelessness, *sapiens* would have resorted to community strategies of prevention and defence, which would involve dexterity, strong ties, and social cohesion.

In a metabolic aspect, considering that slow postnatal maturation affects the probability of viability of the child, metabolic correlates are not unrelated to the probability of early postnatal survival, which, under normal conditions, results in a higher reserve load during the infant period. However, this results in comparatively lower muscle mass load. Regarding the development of an early feeding strategy – a nutritional and cultural issue – perhaps it is opportune to briefly mention comparative observations elaborated by Leonard et al. (2003) related to body fat content at birth, on which the human species surpasses all the other species studied. This data makes it possible to speculate about the possible "brain-protective" value of the initial energy reserve required by a newborn with a prolonged lack of self-sufficiency and with high metabolic requirements imposed by brain development and comparatively high metabolic demand. Besides the impact of this resource during *sapiens* development, this reserve would not be necessary in species where evolutionary pressure resulted in the emergence of other early postnatal characteristics, such as rapid growth, autonomy, neuromuscular development, and digestive efficiency. An additional observation is that, according to Leonard et al. (2003), the greatest accumulation of fat reserve in the human developmental period coincides with the usual timing of weaning and the passage to a nonmaternal diet of evolutionary value.

For this reason, some current physiological profiles are related to strategies that have been useful under primitive conditions for offspring survival. In some respects, our construction carries inertial components of past evolutionary adaptations. Hence, it can be interpreted that this fat reserve at an early age would have acted as a "cushion" for an adaptive interface to more demanding feeding conditions and to less protected conditions than those that theoretically could be provided today. The result of this lack or decrease in energy reserve is that, due to disease or socioeconomic conditions, there is a potential compromise of survival or optimal body and brain development in the face of any additional demand. At present, the risk of early death associated with cultural malnutrition or malnutrition due to poverty, today the menace for child predation is not due to carnivorous felines but to socioeconomic policies. This compounds the risk of compromising the individual's relative functional capacity, impacting its cognitive development, social integration, and job competitiveness.

On organic neurobiological domains, neurogenesis and gliogenesis determine the numerical components of processing units but not the dynamic connectivity among them, which basically depend on microstructural events such as synaptogenesis and sinaptolysis, neurotransmitters, receptors, molecular and ionic transporters, and activity in the neural nets. Cell production and connectivity are one of the central factors in the neocortical volume and gyrification process. Developmental neurogenesis (Rakic 1995a, 1995b; Rash et al. 2016) occurs in a subventricular region during prenatal development. Hence, gestation duration and the number of cortical cell duplication cycles are critical factors in the neurogenetic equation. During such a period, the number of cycles of cell generation contributes to the initial cellular brain content, which is followed by a process of postnatal pruning through neuronolysis and synaptolysis processes, conditioned to environmental variables that include exposure to cognitive and nutritional that also affects gut microbiome development.

The intense brain cell production period that lasts until about the second year of life is followed by regressive physiological processes that eliminate excess cells and connections. The system becomes tighter and more efficient by ridding itself of unnecessary units. However, in this case, what is "necessary" is defined not only by genetic programs or humoral messages (from the internal environment) but also by exposure to stimuli and demands from the external environment. These are translated into phenomena at the brain cell level, whether as activation or inhibition. The result is the stabilisation or reconfiguration of neural circuits, either by suppression or survival of connectedness. In fact, the claim that activity is the key to survival has applications at many levels of the human living conditions, involving the organisation of the nervous system and mental abilities. A series of structural (cellular processes and their contacts) and functional (neural dynamic characteristics) parameters of the cerebral cortex and the organisation of behaviour are sensitive to postnatal rearing conditions. That is, the expression of the genome programs (Black 1998; Plomin and Kosslyn 2001; Evans et al. 2005) would depend on signals provided by the environment, whether internal via the extracellular medium during the period of intrauterine development or from the external, egocentric environment, both physical and affective, during the period of extrauterine life.

On comparative grounds, applying mathematical models, Lewitus et al. (2014) showed variations in the proliferative potential of the subventricular region among mammalian species, a mechanistic determinant of neocortical expansion. According to those authors, species with low proliferative index – expressed in a lesser cortical folding, or *lysencephalia* – tend to live in comparatively smaller habitats and reduced social groups, compared to those species with a larger folding index – or *gyrencephalia* –, a characteristic with positive evolutive pressure, expressed in specific mammalian lineages (Weaver 2014).

The neurobiological scaffolding for generating and expressing our species profile might be affected by environmental conditions – most definitely nutritional factors and host-gut microbiome set-up – in which individual development occurs. These affect primarily microstructural brain events and processes rather than macrostructural ones.

In humans, the prolonged postnatal development of the brain has, among others, two functional consequences: the possibility of continuing brain growth beyond the intrauterine period (Jernigan et al. 2011; Raznahan et al, 2012) and the opportunity for early environmental impacts (familiar or not) to affect the organisation of biological characteristics. This opportunity to essentially continue brain development under extrauterine conditions – one of the evolutionary circumstances that mark our species – is a permissive condition for developing individual profile on cognitive and emotional complexity. This is possible due to the characteristics of "plasticity" of the brain organisation and behaviour. However, this continued process imposes a continuous nutritional demand mediated by host-gut microbiome interactions, which are further considered in successive chapters.

Although adaptability is a characteristic of living organisms in general, neuroplasticity is an intrinsic property of the nervous system, which is maintained to a greater or lesser extent throughout life and on which its complex functioning depends. During the early stages of development, it acquires particular significance to the extent that it underlies the temporal and spatial configuration processes of brain organisation. Such processes constitute the biological substratum of our representations of the external and internal world and our interaction with them, the development of cognitive abilities and emotional processing, and the organisation of behaviour based on experience. This postnatal opportunity for the prolonged cultural or social moulding of the brain and mental activity requires special nutritional and social care for the conditions of its development until adolescence. During the different stages of development, the environment will leave traces of variable significance for moulding individual phenotypes.

For these reasons, the development of brain functioning, and mental activity characteristics are not phenomena marginalised from the physical and social influences of the environment. However, the consequences of the latter will vary depending on the developmental stage in which they act. The brain will complete the initial stages of its organisation in a sub-optimal way in a social environment affected by nutritional deficiencies – due to inadequate availability or interfering disease liabilities – or by toxic substances, and in an external environment characterised by relative isolation and lack of physical and affective stimuli. It could undergo transient or persistent altered sub-optimal developmental processes, both in the intrinsic connective dimension and in cognitive and emotional domains. In short, this condition represents a *social brain damage*; its characteristics will depend on the development stage and the nature and duration of the unsuitable rearing conditions.

The probability and extent that our individual genetic "programs" can be executed "optimally" varies among individuals, be it for reasons of the genome or for circumstantial reasons – whether of the internal or external environment – in which we develop and act. Human sensory receptors and channels do not have absolute identical thresholds, latency times, or response speeds, nor do similarly associate stimuli and general behavioural experiences or store them as identical contents. In addition, the agility and multiplicity of possible verbal and non-verbal representations of the external and intimate world condition human behaviour and ability to analyse and intervene in the environment. With language development, humans contribute to defining the complex internal–dynamic representation of a world parallel to the physical one and its interactions with it. The nature of those representations and the relationships humans build will depend on the timing and conditions in which neuro-behavioural processes operate.

During the initial developmental stages, from conception to approximately two to three years of life, phenomena related to the growth of brain cell populations and their basic connections occur. These will constitute the initial substratum of the adult brain and its operating, conditioning, framework, abilities, emotional profile, and social behaviour. Therefore, the alterations produced during that time –depending on their nature, intensity and duration– will affect individual brain and mental profiles. During this period, alterations in brain-gut microbiome interactions, severe malnutrition, poisoning, deficiencies, infections, maternal consumption of legal or prohibited toxic substances, and parental inattention can constitute sources of disturbances in brain organisation and its mental and behavioural expression, as well as on gut microbiome composition.

From the third year of postnatal life, brain changes are more closely linked to the production, facilitation, and inhibition of new connections and circuit pathways, a biological substrate for the organisation of cognitive and emotional mental processes and language development. For this reason, the consequences of alterations in this period can have a more hidden expression, and the cognitive damage is manifested in performance in the face of greater demands.

As a species, humans must fully assume that its comparative advantage lies in this extended period of postnatal brain immaturity since with it also comes a prolonged phase of brain development, social "shaping" of the mind, and brain-gut microbiome interactions. However, despite those fundamental principles for survival – those related to childcare –, numerous modern communities survive today in conditions that contradict these basic requirements, thus condemning child populations to insufficient conditions and inadequate nutrition, health, and social support (Turkheimer et al. 2003). This potentially causes social damage to the infant brain, its persistence over time, and the recipient's age. In the profound inequalities of child upbringing lies the germ of profound differences in individual aptitudes and performance, affecting social mobility.

*

In their evolutionary journey, human communities built imaginary worlds and productive, emotional, and aesthetic structures that generated a rich variety of proposals. These have allowed them to survive in all corners of the world and develop varied individual profiles since their ancestors abandoned the niche of the African savanna. This plastic, imaginative behaviour increased the number of possible alternatives for the species' survival. In fact, the phenotypic and cultural variety is an inseparable part of the central engine of the adaptive capacity of *Homo sapiens*. From ancient nomadic or sedentary communities to complex feudal, autocratic, democratic, pagan, agnostic, mystical, religious, socialist, capitalist, federative, colonialist, imperialist, matriarchal or patriarchal organisations, the diversity of individual talents, beliefs or forms of social organisation that characterise the history of human civilisation would not have been expressed if the former had not existed and the variety of cultural contexts had not prevailed.

Above all, this variety implies a range of initiatives and adaptability span of the species. It contrasts with the notion of a unique social model or a single, paradigmatic stereotypic thought, whether referred to individuals or a community. This makes cultural uniformity an enemy of our species' development and future. In this context, the suppression of cultural variety and the freedom to develop communities with alternative organisational offers would result in a threat to the species, representing an immoral counter-evolutionary action since it tends to obliterate one of the primary characteristics for the survival of our species: its ability to develop a rich menu of possible strategies and interindividual variations and phenotypes. It is deeply immoral insofar as it puts the species' future at risk, a transcendent ethical value if there is one. To avoid this, socially alert populations are needed, with the spark of creativity applied to every contingency in life, no matter how small. In this context, the deterioration of the child's full capacity for brain and mental development by policies favouring or generating conditions of deep social inequity and cognitive and nutritional deficiencies are part of such evolutive immorality.

*

Experimental Research on Brain *Critical Period*

Brain plasticity is maximal at specific time windows during early development, known as the *critical period*. The basic concept of *critical period* within neurodevelopmental domains takes into consideration the original experiments by Hubel and Wiesel (1963) following early monocular deprivation experiments, who stated that,

... These experiments, then, indicate that very young kittens are particularly susceptible to the effects of deprivation, and that the susceptibility decreases with each month of life, possibly even vanishing in the mature animal.

The extended period during early childhood brain development represents a crucial stage during which a series of variables could act positively or negatively. The transition from a plastic to a more fixed state – in terms of behavioural processing profiles – allows for the sequential consolidation and retention of new and more complex perceptual, motor, and cognitive functions. Developmental alterations during critical periods have been implicated in several neurocognitive profiles and disorders. These processes are not germane to gut microbiome development and composition, thus involving host-gut microbiome interactions.

As discussed in Montiel-Castro et al. (2013), the number of microorganisms inhabiting the human body has impacted self-perception, from a self-sufficient individual to the perception of our bodies as super-complex ecosystems. This perspective change included a reappraisal of the role of microorganisms within our bodies. Across evolution, as microbial life was increasingly tolerated, endosymbionts have established important feedback channels with the CNS, some of which are crucial for maintaining homeostasis. To what extent does this symbiotic construction impact brain development's critical periods?

In terms of the gut microbiome, as discussed in Amato (2016), because *Lactobacillus* dominates the human vaginal tract, it is one of the first microbes to colonise the infant gut. Different exposure patterns to Lactobacilli species and strains are likely to lead to distinct developmental trajectories in CNS function, especially if exposure occurs in early childhood when gut microbial impacts on the brain are most substantial. Amato (2016) considered that host-gut microbe interactions would shape host plasticity and fitness in various contexts (nutrition, health, and behaviour) and, therefore, represent a key factor missing from existing models of human and nonhuman primate ecology and evolution. In further studies, Amato et al. (2017) analysed the effect of host kinship and time spent in social contact among the non-human primate black howler monkeys (*Alouatta nigra*). They observed that social interactions are associated with variation in gut microbiota composition, even in arboreal primates that live in small social groups and spend a relatively low proportion of their time in physical contact. It should be stressed that, as noted by Foster and McVey Neufeld (2013), alterations in microbiota would modulate plasticity-related serotonergic and GABAergic signalling systems in the CNS.

Perhaps one should take note that some behavioural adaptations are related – as inertial or remnant – to strategies useful for evolutive primitive conditions of survival of the offspring. Hence, in some domains we are probably built to implement components of past evolutionary conditions.

The basic instalment of individual development, self-construction, and behavioural performance depends on genetic, physiological, and interactive environmental cues during pre-and postnatal environments. Postnatally, the potential self becomes an additional element in an interactive personal construction. What follows are some considerations on what are considered critical or sensitive periods within neuroplasticity. Though the latter represents a native property of the neural organisation that persists throughout life with differing properties, early developmental instances provide the basis for instilling persistent physiological and behavioural profiles.

In 1949, Donald Hebb published his key concept based on use-dependent neural plasticity, which would later influence neuroscience developmental domains. Within brain domains, a series of structural (neural cell proliferation and connectivity) and functional (dynamic characteristics of neural elements and assemblies) parameters of the cerebral cortex and the organisation of open behaviour are sensitive to preterm and rearing conditions. The neuro-behavioural expression of the genome programs depends on signals provided by the environment, whether internal via the extracellular medium during intrauterine development or from the external, egocentric environment, both physical and affective, during extrauterine life. Neural plasticity is modified by experience and environmental stimuli and continues to play a role in synaptic remodelling in the developing and adult nervous systems, as related to processes involving adult neurogenesis. In this regard, Ge et al. (2007) reported that adult-born neurons – observed in the hippocampus – exhibit classic critical period plasticity as neurons in the developing nervous system. Its transient nature may provide a fundamental mechanism supporting experience-induced plasticity and the production of cohorts of new neurons exhibiting transient enhanced plasticity, thus potentially expanding the capacity of the adult circuitry to be modified by experience throughout life.

Based on previous reports (as quoted below), Knudsen (2004) stressed that early experience stimuli may exert a long-lasting influence on the structural and dynamic development of brain circuits, sometimes persistent enough to consider that it defines a *critical period* in development, considered as a *sensitive period*. As summarised by Dombrovski and Condron (2021), many sensory processing regions of the central brain undergo critical/sensitive periods of experience-dependent plasticity. During this time, ethologically relevant information shapes circuit structure and function. Data on the impact of social and hormonal variables on the organisation of the central nervous system and predominant experience during the early developmental period largely relies on experimental work performed in rodents. Cowan et al. (2020) considered physiologically relevant that periods of change in the microbiota coincide with the development of other body systems, particularly the brain. Disruption of such sequence at specific developmental windows would affect specific regulatory functions. In this regard, on behavioural grounds, Christian et al. (2015), Aatsinki et al. (2019) and Fox et al. (2021) discussed the early sensitive period related to microbial gut colonisation and its impact on child temperament traits related to negative affectivity and surgency/extraversion, and its relationship for developing mental health outcomes.

Though species development differences must be considered, the following reports provide a set of contingencies that in other species would take place at other developmental periods.

There is general recognition that the developing nervous system is qualitatively different from the adult nervous system, as stated by Rice and Barone (2000), and express developmental stages of increased neurobehavioural plasticity. In this regard early postnatal behavioural experience and behavioural imprinting share a common result linked with the conditions for neural plasticity. In this domain, the work by Nikolas Tinbergen (1907–1988) and Conrad Lorenz (1903–1989) – founders of modern ethology – provided initial grounds to describe such behaviours. In non-human primates, the initial experimental work by Harlow (1959) on minimal physical components gating maternal adherence behaviour by infantile primates added to such a current concept. The impact of early environmental enrichment and complexity on neurobehavioural development was tackled by a series of rodent experiments by the groups of Greenough, Rosenzweig, and Bennett (Krech et al. 1960; Rosenzweig et al. 1962; Altman and Das 1964; Bennett et al. 1964; Rosenzweig 1966; Rosenzweig et al. 1972a, b; Volkmar and Greenough 1972; Greenough and Volkmar 1973; Greenough et al. 1973), to which should be included additional studies. Among others, in particular those of Hubel and Wiesel (1963) on cat visual system following monocular visual deprivation; of Turner and Greenough (1985) on differential rearing effects on rat visual cortex; of Winterfeld et al. (1998) on the impact of social environment on working memory in gerbils; of Rampon et al. (2000) on gene expression following enrichment training, reviewed by Diamond (2001) on the impact of environmental enrichment on cerebral cortex microstructure; and on synaptic plasticity by Kolb et al. (2003). According to data obtained – mostly in rodents – by these and other authors, it has been confirmed that, the environment's physical conditions and social factors during the rearing period affect brain development at the microstructural and neuro-chemical level, as well as profiling the adult's cognitive abilities and emotional behaviour.

In chimpanzees, Davenport et al. (1973) reported the impact of early experience of impoverished conditions on cognitive development as evaluated later when adults. Those submitted to restricted laboratory environments showed impaired performance on object quality discrimination compared with animals born in the wild. Kozor-ovitskiy et al. (2005) reported that environmental complexity influences the structure and biochemistry of marmoset brains. These changes occurred following one month of living in a more complex environment. As described by the authors, they essentially consisted in increased dendritic spine density on dentate gyrus granule cells, hippo-campal field CA1 pyramidal cells, and prefrontal cortex pyramidal cells. These authors further reported that increased levels of the presynaptic protein synaptophysin – an integral membrane protein of synaptic vesicles – were detected in the mentioned brain structures of the marmosets. Probable basic mechanisms underlying some of these critical periods were further studied in early partial visual deprivation cases. As reported by Mowery and Garraghty (2022), membrane expression levels of receptor subunit mapping came to reflect those seen in early phases of critical period development. According to the authors, these observations suggested a developmental recapitulation since, following prolonged sensory deprivation, the adult cells returned to a critical period, like in plastic states. Environmental enrichment paradigms prolong and enhance neural plasticity. Experimental studies on nonhuman animals demonstrate that environmental stimulation, parental nurturance, and early life stress affect brain structure and functioning. Regarding neurophysiological involvement, Herringa et al. (2016) proposed that frontal-amygdaline connectivity would be involved in early adaptive mechanisms towards adversity. Adaptation in the face of potentially stressful challenges involves the activation of neural, neuroendocrine, and neuroendocrine-immune mechanisms (*allostasis*, McEwen 1998).

In humans, Chugani (1998) referred to critical periods based on brain glucose uti-lisation monitored with Positron Emission Tomography. These periods of high glucose utilisation underscore postnatal neural maturation at different postnatal times across brain structures. As stated by the author, the highest degree of glucose metabolism in the newborn is in the primary sensory and motor cortex, cingulate cortex, thalamus, brain stem, cerebellar vermis, and hippocampal region. At two to three months of age, glucose utilisation increases in the parietal, temporal, and primary visual cortex, basal ganglia, and cerebellar hemispheres. Between six and 12 months, glucose utilisation increases in the frontal cortex. These high rates of glucose utilisation would only subside past year 10, while the cerebral cortex would undergo a dynamic course of metabolic maturation that persists until ages 16–18 years. These patterns of glucose utilisation would be associated with functional changes in the different brain neural regions and the emergence of various behaviours.

*

Early Experience and Human Development

The critical role of social interactions in driving phenotypic variation has long been inferred from the association between early social deprivation and adverse neurodevelopmental outcomes involving epigenetic mechanisms (Champagne 2010), which include early impact on gut microbiome development. In this regard, poverty and indigence place a profound, urgent, and ethical demand, considering that lack of long-term public social policies could hamper recovery when such conditions unduly persist in time (Colombo 2007).

Genome expression is regulated by the epigenome (the set of chemical modifications to the DNA and DNA-associated proteins in the cell, which conditions gene expression). The environmental components affect the epigenome, while the epigenetic and genetic individual makeup modulates the response to them. As considered by Szyf et al. (2008), two components of the epigenome are chromatin structure and covalent modification of the DNA molecule by methylation sculpted during development. According to the authors, DNA methylation is dynamic later in life in postmitotic cells such as neurons and thus potentially responsive to different environmental stimuli throughout life, thus building the epigenetic individual profiles. DNA methylation has been associated with several processes, including cognitive disabilities (Pogribny and Beland 2009).

Behavioural studies are providing new views on the relationship between the social environment and epigenetic programming. The human prefrontal cortex plays a critical role in human cognitive processes and shows a protracted postnatal maturation period. Numata et al. (2012) studied DNA methylation in the development of the dorsolateral prefrontal cortex across the lifespan, based on human brains from non-psychiatric controls collected at the Clinical Brain Disorders Branch (National Institute of Mental Health). According to the authors, at the genome level, the transition from foetal to postnatal life is characterised by a reversal of direction, from prenatal demethylation to increased postnatal methylation. The fastest changes occur during the prenatal period, which slows down markedly after birth and continues to further slow down with ageing. Marioni et al. (2018) demonstrated a link between blood-based DNA methylation and measures of phonemic verbal fluency and global cognitive ability, suggesting that blood-based methylation signatures may be useful tools to explore differences in brain-related outcomes. Additionally, as shown by Swann et al. (2020), in the mouse brain, microbial-associated molecules can cross the blood brain barrier and bind receptors expressed in the brain.

*

Socioeconomic disparities are associated with differences in cognitive development, calling attention to its multifactorial impact on cognitive development, including the gut microbiome environment. Early social and environmental conditions in humans impact cognitive performance, as discussed in Lipina et al. (2005), Diamond et al. (2007), Colombo (2007), Lipina and Colombo (2009), Lipina et al. (2013), among other reports. Consequently, children living in poverty generally perform poorly in school, with markedly lower standardised test scores and educational attainment, as Hair et al. (2015) stated. The longer children live in poverty, the greater their academic deficits. These patterns persist to adulthood, contributing to lifetime-reduced occupational attainment. According to these authors, on average, children from low-income households scored four to seven points lower on standardised tests, and as much as 20% of the gap in test scores could be explained by neural maturational lags in the frontal and temporal lobes. As stated by Hackman et al. (2015), childhood socio-economic status (SES) (a multidimensional construct) predicts executive function (EF) (it includes basic cognitive processes such as attentional control, cognitive inhibition, inhibitory control, working memory, and cognitive flexibility) (reviewed by Hackman and Farah, 2009). According to these authors, SES is an important predictor of neurocognitive performance, particularly of language and executive function.

> The relevance of SES to cognitive neuroscience lies in its surprisingly strong relationship to cognitive ability as measured by IQ and school achievement beginning in early childhood.
>
> *(Hackman and Farah 2009)*

Noble et al. (2015) administered a standardised structural magnetic resonance imaging (MRI) protocol to investigate relationships between socioeconomic factors (parent education, family income) and brain morphometry (cortical surface area). Relationships were most prominent in regions supporting language, reading, executive functions, and spatial skills; surface area mediated socioeconomic differences in certain neurocognitive abilities. The authors concluded that parental education and family income account for individual variation in independent characteristics of brain structural development in regions critical for developing language, executive functions and memory. Applying similar magnetic resonance procedures, Hair et al. (2015) concluded that children from families with limited financial resources displayed systematic structural differences in the frontal lobe, temporal lobe, and hippocampus. Thus, specific brain structures tied to processes critical for learning and educational functioning are vulnerable to the environmental circumstances imposed by poverty. Working memory (WM) capacity reflects executive functions associated with performance on various cognitive tasks and education outcomes. In this regard, Finn et al. (2017)

reported variability of working memory (WM) in adolescents depending on family income. The finding that lower income was associated with lesser WM capacity is consistent with prior findings that lower-SES children and adolescents perform worse on measures of executive function, including spatial working memory and verbal tasks (Farah et al. 2006; Hackman et al. 2015; Heckman 2007; Noble et al. 2015). These authors concluded that there is clear evidence that WM capacity is an essential determinant of cognitive ability and that children from lower-socioeconomic status (SES) environments, relative to higher-SES environments, perform worse on many measures of cognitive ability, including working memory capacity. As stated in Hackman et al. (2015), early emerging and persistent SES-related differences in executive function, partially explained by characteristics of the home and family environment – including feeding habits and food access –, represent a potential source of socioeconomic disparities in achievement and health across development.

Last et al. (2018) examined the association between childhood SES and executive function (EF) in a USA sample ranging from 9–25 years of age and found positive relations between both variables, with no change by age. These studies suggest that the SES disparity in EF is established early in life and holds into early adulthood. Chan et al. (2018) included studies on functional interactions between areas of the brain as measured at rest (resting-state functional correlations, RSFCs) in middle-aged adults (35–64 years). The RSFCs of brain areas are organised into large-scale brain networks that consist of multiple segregated subnetworks or modules. RSFC system segregation and cortical grey matter thickness exhibit systematic differences across adult age. Middle-aged participants with lower SES exhibited reduced segregation of the systems in their large-scale functional brain networks and thinner mean cortical grey matter compared with higher SES individuals in an equivalent age range. Lower SES individuals exhibit less organised functional brain networks and reduced cortical thickness than higher SES individuals. These studies should be coupled with those aiming at the socioeconomic impact on early feeding habits and the development and maintenance impact on the gut microbiome. As reported by Tooley et al. (2021), children growing up in higher socioeconomic status homes tend to be exposed to more complex and cognitively stimulating environments, and cognitive enrichment is associated with improved cognition.

Though animal species undergo different rates of brain organisation, which condition the grade and domain of neural plasticity and response to environmental cues, previous comments address the issue of early developmental conditions on several social and cognitive domains, which affect individual performance and social insertion in adolescent and adult life. Additionally, as stated by Seebacher and Krause (2019), the environment affects the physiology of individuals via epigenetic mechanisms, and individual physiology influences conspecific interactions; at a higher level of organisation, these conspecific interactions could scale up to social domains.

The above environmental conditioners on early human development are further enriched by diet and eating habits that impact the instalment and development of the gut microbiome.

References

Aatsinki, Anna-Katariina, *et al.* "Reply to the Letter to the Editor: Gut Microbiota Composition is Associated with Temperament Traits in Infants." *Brain, Behaviour, and Immunity*, vol. 81, 2019, pp. 671–672, doi:10.1016/j.bbi.2019.07.006.

Altman, Joseph, and Gopal D. Das. "Autoradiographic Examination of the Effects of Enriched Environment on the Rate of Glial Multiplication in the Adult Rat Brain." *Nature*, vol. 204, no. 4964, 1964, pp. 1161–1163, doi:10.1038/2041161a0.

Amato, Katherine R. "Incorporating the Gut Microbiota into Models of Human and Non-Human Primate Ecology and Evolution." *American Journal of Physical Anthropology*, vol. 159, no. S61, 2016, pp. 196–215, doi:10.1002/ajpa.22908.

Amato, Katherine R., *et al.* "Patterns in Gut Microbiota Similarity Associated with Degree of Sociality among Sex Classes of a Neotropical Primate." *Microbial Ecology*, vol. 74, no. 1, 2017, pp. 250–258, doi:10.1007/s00248-017-0938-6.

Bennett, Edward L., *et al.* "Chemical and Anatomical Plasticity of Brain." *Science*, vol. 146, no. 3644, 1964, pp. 610–619, doi:10.1126/science.146.3644.610.

Black, Ira B. "Genes, Brain, and Mind: The Evolution of Cognition." *Neuron*, vol. 20, no. 6, 1998, pp. 1073–1080, doi:10.1016/S0896-6273(00)80489-4.

Champagne, Frances A. "Epigenetic Influence of Social Experiences Across the Lifespan." *Developmental Psychobiology*, vol. 52, no. 4, 2010, pp. 299–311, doi:10.1002/dev.20436.

Chan, Micaela Y., *et al.* "Socioeconomic Status Moderates Age-Related Differences in the Brain's Functional Network Organization and Anatomy across the Adult Lifespan." *Proceedings of the National Academy of Sciences*, vol. 115, no. 22, 2018, pp. E5144–E5153, www.pnas.org/content/115/22/E5144, doi:10.1073/pnas.1714021115.

Christian, Lisa M., *et al.* "Gut Microbiome Composition is Associated with Temperament during Early Childhood." *Brain, Behavior, and Immunity*, vol. 45, 2015, pp. 118–127, www.ncbi.nlm.nih.gov/pmc/articles/PMC4342262/, doi:10.1016/j.bbi.2014.10.018.

Chugani, Harry T. "A Critical Period of Brain Development: Studies of Cerebral Glucose Utilization with PET." *Preventive Medicine*, vol. 27, no. 2, 1998, pp. 184–188, www.sciencedirect.com/science/article/pii/S0091743598902742, doi:10.1006/pmed.1998.0274.

Colombo, Jorge A. *Pobreza y Desarrollo Infantil. Una Contribucion Multidisciplinaria*. Buenos Aires: Ediciones Paidós, 2007, pp. 97–113.

Cowan, Caitlin S. M., *et al.* "Annual Research Review: Critical Windows – the Microbiota–Gut–Brain Axis in Neurocognitive Development." *Journal of Child Psychology and Psychiatry*, vol. 61, no. 3, 2020, pp. 353–371, doi:10.1111/jcpp.13156.

Davenport, Richard K., *et al.* "Long-Term Cognitive Deficits in Chimpanzees Associated with Early Impoverished Rearing." *Developmental Psychology*, vol. 9, no. 3, 1973, pp. 343–347, doi:10.1037/h0034877.

Diamond, A., *et al.* "Preschool Program Improves Cognitive Control." *Science*, vol. 318, no. 5855, 2007, pp. 1387–1388, doi:10.1126/science.1151148.

Diamond, Marian C. "Response of the Brain to Enrichment." *Anais Da Academia Brasileira de Ciências*, vol. 73, no. 2, 2001, pp. 211–220, doi:10.1590/s0001-37652001000200006.

Dombrovski, Mark, and Barry Condron. "Critical Periods Shaping the Social Brain: A Perspective from *Drosophila*." *BioEssays*, vol. 43, no. 1, 2021, doi:10.1002/bies.202000246.

Evans, P. D., *et al.* "*Microcephalin*, a Gene Regulating Brain Size, Continues to Evolve Adaptively in Humans." *Science*, vol. 309, no. 5741, 2005, pp. 1717–1720, doi:10.1126/science.1113722.

Farah, Martha J., *et al.* "Childhood Poverty: Specific Associations with Neurocognitive Development." *Brain Research*, vol. 1110, no. 1, 2006, pp. 166–174.

Finn, Amy S., *et al.* "Functional Brain Organization of Working Memory in Adolescents Varies in Relation to Family Income and Academic Achievement." *Developmental Science*, vol. 20, no. 5, 2017, p. e12450, doi:10.1111/desc.12450.

Foster, Jane A., and Karen-Anne McVey Neufeld. "Gut–Brain Axis: How the Microbiome Influences Anxiety and Depression." *Trends in Neurosciences*, vol. 36, no. 5, 2013, pp. 305–312.

Fox, Molly, *et al.* "Development of the Infant Gut Microbiome Predicts Temperament across the First Year of Life." *Development and Psychopathology*, vol. 34, no. 5, 2021, pp. 1–12, doi:10.1017/s0954579421000456.

Ge, Shaoyu, *et al.* "A Critical Period for Enhanced Synaptic Plasticity in Newly Generated Neurons of the Adult Brain." *Neuron*, vol. 54, no. 4, 2007, pp. 559–566, doi:10.1016/j.neuron.2007.05.002.

Greenough, William T., *et al.* "Effects of Rearing Complexity on Dendritic Branching in Frontolateral and Temporal Cortex of the Rat." *Experimental Neurology*, vol. 41, no. 2, 1973, pp. 371–378. doi:10.1016/0014-4886(73)90278-1.

Greenough, William T., and Fred R. Volkmar. "Pattern of Dendritic Branching in Occipital Cortex of Rats Reared in Complex Environments." *Experimental Neurology*, vol. 40, no. 2, 1973, pp. 491–504. doi:10.1016/0014-4886(73)90090-3.

Hackman, Daniel A., and Martha J. Farah. "Socioeconomic Status and the Developing Brain." *Trends in Cognitive Sciences*, vol. 13, no. 2, 2009, pp. 65–73, www.ncbi.nlm.nih.gov/pmc/articles/PMC3575682/, doi:10.1016/j.tics.2008.11.003.

Hackman, Daniel A., *et al.* "Socioeconomic Status and Executive Function: Developmental Trajectories and Mediation." *Developmental Science*, vol. 18, no. 5, 2015, pp. 686–702, doi:10.1111/desc.12246.

Hair, Nicole L., *et al.* "Association of Child Poverty, Brain Development, and Academic Achievement." *JAMA Pediatrics*, vol. 169, no. 9, 2015, pp. 822–829, jamanetwork.com/journals/jamapediatrics/article-abstract/2381542, doi:10.1001/jamapediatrics.2015.1475.

Hebb, D. O. *The Organization of Behaviour: A Neuropsychological Theory.* New York: John Wiley & Sons, 1949.

Heckman, James J. "The Economics, Technology, and Neuroscience of Human Capability Formation." *Proceedings of the National Academy of Sciences*, vol. 104, no. 33, 2007, pp. 13250–13255, doi:10.1073/pnas.0701362104.

Herringa, Ryan J., *et al.* "Enhanced Prefrontal-Amygdala Connectivity Following Childhood Adversity as a Protective Mechanism against Internalizing in Adolescence." *Biological Psychiatry: Cognitive Neuroscience and Neuroimaging*, vol. 1, no. 4, 2016, pp. 326–334, doi:10.1016/j.bpsc.2016.03.003.

Hubel, David H., and Torsten N. Wiesel. "Receptive Fields of Cells in Striate Cortex of Very Young, Visually Inexperienced Kittens." *Journal of Neurophysiology*, vol. 26, no. 6, 1963, pp. 994–1002, doi:10.1152/jn.1963.26.6.994.

Jernigan, Terry L., *et al.* "Postnatal Brain Development: Structural Imaging of Dynamic Neurodevelopmental Processes." *Progress in Brain Research*, vol. 189, 2011, pp. 77–92, www.ncbi.nlm.nih.gov/pmc/articles/PMC3690327/, doi:10.1016/B978-0-444-53884-0.00019-1.

Kerhoas, Daphne, *et al.* "Social and Ecological Factors Influencing Offspring Survival in Wild Macaques." *Behavioral Ecology*, vol. 25, no. 5, 2014, pp. 1164–1172, doi:10.1093/beheco/aru099.

Knudsen, Eric I. "Sensitive Periods in the Development of the Brain and Behavior." *Journal of Cognitive Neuroscience*, vol. 16, no. 8, 2004, pp. 1412–1425, doi:10.1162/0898929042304796.

Kohler, Timothy A., *et al.* "Greater Post-Neolithic Wealth Disparities in Eurasia than in North America and Mesoamerica." *Nature*, vol. 551, no. 7682, 2017, pp. 619–622, www.nature.com/articles/nature24646, 10.1038/nature24646.

Kolb, Bryan, *et al.* "Experience-Dependent Changes in Dendritic Arbor and Spine Density in Neocortex Vary Qualitatively with Age and Sex." *Neurobiology of Learning and Memory*, vol. 79, no. 1, 2003, pp. 1–10, doi:10.1016/s1074-7427(02)00021-7.

Kozorovitskiy, Y., *et al.* "Experience Induces Structural and Biochemical Changes in the Adult Primate Brain." *Proceedings of the National Academy of Sciences*, vol. 102, no. 48, 2005, pp. 17478–17482, doi:10.1073/pnas.0508817102.

Krech, D., *et al.* "Effects of Environmental Complexity and Training on Brain Chemistry." *Journal of Comparative and Physiological Psychology*, vol. 53, no. 6, 1960, pp. 509–519, doi:10.1037/h0045402.

Last, Briana S., *et al.* "Childhood Socioeconomic Status and Executive Function in Childhood and Beyond." *PloS One*, vol. 13, no. 8, 2018, p. e0202964, doi:10.1371/journal.pone.0202964.

Leonard, William R., *et al.* "Metabolic Correlates of Hominid Brain Evolution." *Comparative Biochemistry and Physiology Part A: Molecular & Integrative Physiology*, vol. 136, no. 1, 2003, pp. 5–15, doi:10.1016/s1095-6433(03)00132-6.

Lewitus, Eric, *et al.* "An Adaptive Threshold in Mammalian Neocortical Evolution." *PLoS Biology*, vol. 12, no. 11, 2014, p. e1002000, doi:10.1371/journal.pbio.1002000.

Lipina, Sebastián J., *et al.* "Performance on the A-not-B Task of Argentinean Infants from Unsatisfied and Satisfied Basic Needs Homes." *Revista Interamericana de Psicología/Interamerican Journal of Psychology*, vol. 39, no. 1, 2005, pp. 49–60, https://www.redalyc.org/articulo.oa?id=28439106.

Lipina, Sebastián J., and Jorge A. Colombo. *Poverty and Brain Development during Childhood: An Approach from Cognitive Psychology and Neuroscience*. Washington, DC: American Psychological Association, 2009.

Lipina, Sebastián J., *et al.* "Linking Childhood Poverty and Cognition: Environmental Mediators of Non-Verbal Executive Control in an Argentine Sample." *Developmental Science*, vol. 16, no. 5, 2013, pp. 697–707. doi:10.1111/desc.12080.

Marioni, Riccardo E., *et al.* "Meta-Analysis of Epigenome-Wide Association Studies of Cognitive Abilities." *Molecular Psychiatry*, vol. 23, no. 11, 2018, pp. 2133–2144, doi:10.1038/s41380-017-0008-y.

McEwen, Bruce. "Stress, Adaptation, and Disease: Allostasis and Allostatic Load." *Annals of the New York Academy of Sciences*, vol. 840, no. 1, 1998, pp. 33–44, doi:10.1111/j.1749-6632.1998.tb09546.x.

Mekel-Bobrov, Nitzan, *et al.* "Ongoing Adaptive Evolution of ASPM, a Brain Size Determinant in Homo sapiens." *Science*, vol. 309, no. 5741, 2005, pp. 1720–1722, doi:10.1126/science.1116815.

Montiel-Castro, Augusto J., *et al.* "The Microbiota-Gut-Brain Axis: Neurobehavioral Correlates, Health and Sociality." *Frontiers in Integrative Neuroscience*, vol. 7, no. 70, 2013, doi:10.3389/fnint.2013.00070.

Mowery, Todd M., and Preston E. Garraghty. "Adult Neuroplasticity Employs Developmental Mechanisms." *Frontiers in Systems Neuroscience*, vol. 16, 2022, p. 1086680, pubmed.ncbi.nlm.nih.gov/36762289/, doi:10.3389/fnsys.2022.1086680.

Noble, Kimberly G., *et al.* "Family Income, Parental Education and Brain Structure in Children and Adolescents." *Nature Neuroscience*, vol. 18, no. 5, 2015, pp. 773–778, doi:10.1038/nn.3983.

Numata, Shusuke, *et al.* "DNA Methylation Signatures in Development and Aging of the Human Prefrontal Cortex." *American Journal of Human Genetics*, vol. 90, no. 2, 2012, pp. 260–272, doi:10.1016/j.ajhg.2011.12.020.

Plomin, Robert, and Stephen M. Kosslyn. "Genes, Brain and Cognition." *Nature Neuroscience*, vol. 4, no. 12, 2001, pp. 1153–1154, doi:10.1038/nn1201-1153.

Pogribny, Igor P., and Frederick A. Beland. "DNA Hypomethylation in the Origin and Pathogenesis of Human Diseases." *Cellular and Molecular Life Sciences*, vol. 66, no. 14, 2009, pp. 2249–2261, doi:10.1007/s00018-009-0015-5.

Rakic, Pasko. "A Small Step for the Cell, a Giant Leap for Mankind: A Hypothesis of Neocortical Expansion during Evolution." *Trends in Neurosciences*, vol. 18, no. 9, 1995a, pp. 383–388, doi:10.1016/0166-2236(95)93934-p.

Rakic, Pasko. "Radial versus Tangential Migration of Neuronal Clones in the Developing Cerebral Cortex." *Proceedings of the National Academy of Sciences*, vol. 92, no. 25, 1995b, pp. 11323–11327, doi:10.1073/pnas.92.25.11323.

Rampon, C., *et al.* "Effects of Environmental Enrichment on Gene Expression in the Brain." *Proceedings of the National Academy of Sciences*, vol. 97, no. 23, 2000, pp. 12880–12884, doi:10.1073/pnas.97.23.12880.

Rash, Brian G., *et al.* "Bidirectional Radial Ca2+ Activity Regulates Neurogenesis and Migration during Early Cortical Column Formation." *Science Advances*, vol. 2, no. 2, 2016, p. e1501733, www.ncbi.nlm.nih.gov/pmc/articles/PMC4771444/, doi:10.1126/sciadv.1501733.

Raznahan, A., *et al.* "Prenatal Growth in Humans and Postnatal Brain Maturation into Late Adolescence." *Proceedings of the National Academy of Sciences*, vol. 109, no. 28, 2012, pp. 11366–11371, doi:10.1073/pnas.1203350109.

Rice, D., and S. Barone. "Critical Periods of Vulnerability for the Developing Nervous System: Evidence from Humans and Animal Models." *Environmental Health Perspectives*, vol. 108, no. suppl. 3, 2000, pp. 511–533, doi:10.1289/ehp.00108s3511.

Rosenzweig, Mark R., *et al.* "Effects of Environmental Complexity and Training on Brain Chemistry and Anatomy: A Replication and Extension." *Journal of Comparative and Physiological Psychology*, vol. 55, no. 4, 1962, pp. 429–437, doi:10.1037/h0041137.

Rosenzweig, Mark R. "Environmental Complexity, Cerebral Change, and Behavior." *American Psychologist*, vol. 21, no. 4, 1966, pp. 321–332, doi:10.1037/h0023555.

Rosenzweig, Mark R., *et al.* "Cerebral Effects of Differential Experience in Hypophysectomized Rats." *Journal of Comparative and Physiological Psychology*, vol. 79, no. 1, 1972a, pp. 56–66, doi:10.1037/h0032527.

Rosenzweig, Mark R., *et al.* "Brain Changes in Response to Experience." *Scientific American*, vol. 226, no. 2, 1972b, pp. 22–29, doi:10.1038/scientificamerican0272-22.

Seebacher, Frank, and Jens Krause. "Epigenetics of Social Behaviour." *Trends in Ecology & Evolution*, vol. 34, no. 9, 2019, pp. 818–830, doi:10.1016/j.tree.2019.04.017.

Swann, Jonathan R., *et al.* "Developmental Signatures of Microbiota-Derived Metabolites in the Mouse Brain." *Metabolites*, vol. 10, no. 5, 2020, p. 172, doi:10.3390/metabo10050172.

Szyf, Moshe, *et al.* "The Social Environment and the Epigenome." *Environmental and Molecular Mutagenesis*, vol. 49, no. 1, 2008, pp. 46–60, doi:10.1002/em.20357.

Thompson, Paul M., *et al.* "Genetic Influences on Brain Structure." *Nature Neuroscience*, vol. 4, no. 12, 2001, pp. 1253–1258, www.nature.com/articles/nn758, doi:10.1038/nn758.

Tooley, Ursula A., *et al.* "Environmental Influences on the Pace of Brain Development." *Nature Reviews Neuroscience*, vol. 22, no. 6, 2021, pp. 372–384, www.nature.com/articles/s41583-021-00457-5, doi:10.1038/s41583-021-00457-5.

Tung, Jenny, *et al.* "Social Environment Is Associated with Gene Regulatory Variation in the Rhesus Macaque Immune System." *Proceedings of the National Academy of Sciences*, vol. 109, no. 17, 2012, pp. 6490–6495, 10.1073/pnas.1202734109.

Turkheimer, Eric, *et al.* "Socioeconomic Status Modifies Heritability of IQ in Young Children." *Psychological Science*, vol. 14, no. 6, 2003, pp. 623–628, doi:10.1046/j.0956-7976.2003.psci1475.x.

Turner, Anita M., and William T. Greenough. "Differential Rearing Effects on Rat Visual Cortex Synapses. I. Synaptic and Neuronal Density and Synapses per Neuron." *Brain Research*, vol. 329, no. 1–2, 1985, pp. 195–203, doi:10.1016/0006-8993(85)90525-6.

Volkmar, Fred R., and William T. Greenough. "Rearing Complexity Affects Branching of Dendrites in the Visual Cortex of the Rat." *Science*, vol. 176, no. 4042, 1972, pp. 1445–1447, doi:10.1126/science.176.4042.1445.

Weaver, Janelle. "How Folded Brains Evolved in Mammals." *PLoS Biology*, vol. 12, no. 11, 2014, p. e1002001, doi:10.1371/journal.pbio.1002001.

Winterfeld, Karl T, *et al.* "Social Environment Alters Both Ontogeny of Dopamine Innervation of the Medial Prefrontal Cortex and Maturation of Working Memory in Gerbils (Meriones Unguiculatus)." *Journal of Neuroscience Research*, vol. 52, no. 2, 1998, pp. 201–209, doi:10.1002/(SICI)1097-4547(19980415)52:2<201::AID-JNR8>3.0.CO;2-e.

10

GUT MICROBIOME, DIET, POVERTY, AND CHILD DEVELOPMENT

In microbes, the ecological expansion into new niches is often facilitated by the uptake of gene sequences from other microbes via horizontal gene transfer. In contrast, multicellular eukaryotes do not take up exogenous DNA as readily as microbes: instead, they form symbiotic associations with microbes that carry the necessary genes, allowing a rapid adaptive extension of their phenotypic capabilities. Host microbe symbiosis is widely distributed within the Eukaryota...

(Walter and Ley 2011)

Despite our generally anthropocentric view of the world, it is the microbial population that dominates life on this planet in global diversity and in numbers. The human body itself serves as a scaffold for a multitude of bacteria, archaea, viruses, and eukaryotic microbes that inhabit discrete anatomical niches and outnumber our own somatic and germ cells by an order of magnitude (Turnbaugh et al. 2007).

(Eloe-Fadrosh and Rasko 2013)

Based on the discussion displayed in previous chapters and specific quotations above, it is apparent that the current human cultural construction of the living world and the full concept of freedom is plagued with flaws and omissions. In this review, perhaps a report on a critical test may be missing, such as a strict correlation between our gut microbiome composition and physiological and mental capabilities with our level of free choice and decision process. However, reported studies on comparative grounds involving the human level are seeded with evidence supporting this self-evaluation: our true freedom is also squeezed by our microbiome partner composition and needs. As stated by Eloe-Fadrosh and Rasko (2013),

Despite our generally anthropocentric view of the world, it is the microbial population that dominates life on this planet...

DOI: 10.4324/9781032698380-10

This statement would be expanded if it could be projected onto experimental designs, exploring further the spread and limits of such dominance at the human level.

*

Poverty and Early Malnourishment Impact on Child Development

As applied to human developmental domains, this series of approaches triggers conceptual consequences summarised in the following paragraphs.

> ...Perhaps individuals with special brains (and minds) are more frequent than suspected. They just may go unnoticed due to sociocultural conditions, or their early potential being cancelled following exposure to unwanted health or child rearing hazards during gestation and/or early childhood, or lack of an adequate child-rearing environment. In this context, brain biological variability would represent an additional characteristic of the human species, which ought to be protected from devastation – as produced by famine or lack of challenging environmental stimuli.
>
> *(Colombo et al. 2006)*

Humans are endowed with an extended period of postnatal immaturity compared to other species, during which organic and mental development is completed. From an evolutive point of view, the necessary rearing period imposed particularly demanding conditions for the emergence of *Homo sapiens*, with the postnatal additional requirements to satisfy the demanding caloric and nutritional demands of a developing brain. This depends on the fact that our brain has a comparatively disproportionate size compared to other mammals, considering weight and body size. As stated previously, during the first months of life, the brain accounts for 70–80% of the total body metabolic activity and, in adult life, around 20–25%. This level of cerebral metabolic demand is characteristic of the *Homo sapiens* and is largely due to our expanded cerebral cortex, which had to fold in on itself within a relatively inextensible skull during prolonged childhood (reviews in Colombo 2007, 2020).

*

The prolonged postnatal development of the brain has, among others, two consequences: the possibility of continuing brain growth beyond the intrauterine period and the opportunity for early environmental impacts (familiar or not) to affect the organisation of its biological and dynamic characteristics. This opportunity to continue brain development in extrauterine conditions – one of the evolutionary circumstances that mark the human species – is permissive for developing human cognitive and emotional complexity. This is possible due to the characteristics of "plasticity" of the brain organisation and behaviour. The concept of "neural plasticity" would have been introduced by the philosopher W. James (1890) (cf., Nieto-Sampedro 1996) to describe the susceptibility to behavioural modification and later incorporated by Ramón y Cajal into his structural description and theories on the nervous system. This concept is applied to characterise the possibility of a broad spectrum of changes at the molecular, cellular, and neural network levels and behaviour.

This extended postnatal period poses a series of demands not always met by social conditions, which implies a potential developmental risk. At present, the risk of reducing the relative functional capacity of the individual is added to the risk of early death, wasting or stunting associated with malnutrition, more often due to poverty. This could compromise its full social integration and the level of labour competitiveness. These considerations have a deep root in socio-political domains that will not be further considered here. As stated in the latest report from UNICEF (2023),

> The Joint Child Malnutrition Estimates (JME) released in 2023 reveal insufficient progress to reach the 2025 World Health Assembly (WHA) global nutrition targets and SDG target. Only about one third of all countries are "on track" to halve the number of children affected by stunting by 2030, and assessment of progress to date not being possible for about one quarter of countries. Even fewer countries are expected to achieve the 2030 target of 3% prevalence for overweight, with just 1 in 6 countries currently "on track". Further, an assessment of progress towards the wasting target is not possible for nearly half of countries.

Furthermore, according to the same source,

> While the 2023 edition of the UNICEF-WHO-World Bank Group Joint Malnutrition Estimates shows that stunting prevalence has been declining since 2000, more than one in five – 148.1 million children under 5 – were stunted in 2022, and at least 45.0 million suffered from wasting at any given point of time in the year. Meanwhile, the number of children under 5 affected by overweight worldwide has increased from 33.0 million in 2000 to 37.0 million in 2022.

As commented by Luby et al. (2013),

> The deleterious effects of poverty on child development have been well established in psychosocial research, with poverty identified as being among the most powerful risk factors for poor developmental outcomes. Children exposed to poverty have poorer cognitive outcomes and school performance, and they are at higher risk for antisocial behaviours and mental disorders.

Furthermore, as stated by these authors, exposure to poverty during early childhood is associated with reduced white matter, cortical grey matter, and hippocampal and amygdala volumes measured at school age/early adolescence, detected by magnetic resonance procedures.

Regarding the impact of socioeconomic status on the heritability of the Intelligence Quotient (IQ), Turkheimer et al. (2003) reported that the Wechsler Intelligence Scale for Children scores were analysed in a sample of seven-year-old twins from the National Collaborative Perinatal Project. Results suggest that in impoverished families, 60% of the variance in IQ is accounted for by the shared environment, and the contribution of genes is close to zero, while the results are almost inverse in affluent families. As stated by Meloni (2014), the growing knowledge on *molecular epigenetics* is driving new concepts on the interaction of sociality and brain/mind to a "social biology" conceptual synthesis.

According to Isaacs et al. (2008), randomised intervention trials carried out in the developing and developed world have confirmed the sensitivity of the developing human brain to nutrition, as demonstrated by long-term effects on cognition. This underlines that early nutrition – and its potential impact on the gut microbiome – may affect the development of specific brain structures of fundamental biological importance. The period of basic brain organisation does not end at the infant age, as Chugani (1998) reported based on brain glucose consumption. From the third year of postnatal life, brain changes are more closely linked to the production, facilitation and inhibition of new connections and neural networks, a biological substrate for the organisation of cognitive and emotional mental processes and language development. For this reason, the consequences of the alterations in this period can have a hidden expression in which cognitive damage is manifested in the performance during specific tests in the face of greater demands.

As previously described, neurodevelopment includes a stage during which physical, emotional, and relational conditions can be critical for normal maturation through an imprinting process. Brain development involves a series of crucial cellular, chemical, and connective events (i.e., proliferation, migration, differentiation, synaptogenesis, myelination, synaptic pruning, and apoptosis). Though neuroplasticity characterises basic processes of the neural system, the establishment of individual neural profiles and control functions traverses a time period known as a *critical period*, following which functional plasticity remains as an adaptive process but with comparatively reduced imprinting capacity. Neuroplasticity includes regulatory, neuroendocrine, behavioural, sensory, and cognitive domains. In this regard, Gao et al. (2015) studied human infants during the first year of life – with special emphasis at six months of age – applying multiple longitudinal resting-state functional magnetic resonance imaging scans every three months to delineate detailed growth trajectories of key functional brain networks. A maturation sequence starting with primary sensorimotor/auditory, vision, then attention/default-mode networks, and finally, executive control networks were observed together with network-specific critical periods of growth. This developmental process was affected by socioeconomic status (SES), as revealed by SES-brain correlations, most salient around six months of age.

Hence, the development of brain functioning and mental activity characteristics are not phenomena marginalised by the physical and social influences of the environment. As stated by Farah (2017), no human brain nor neurocognitive development exists outside of a particular socioeconomic context. However, the consequences of the latter will vary depending on the moment of development in which they act. In an internal environment affected by nutritional deficiencies or toxic substances and in an external environment characterised by relative isolation and lack of cognitive, physical, and affective stimuli, the brain will complete the initial stages of its organisation in a sub-optimal way. Therefore, its functioning will have a greater probability of persistent disabilities or modified developmental characteristics, which include cognitive and emotional processing. In short, this configures a potential *social brain damage*. Its characteristics will depend on the developmental stage and the nature and duration of the unsuitable rearing conditions.

Though the concept of *critical periods* has been delineated in previous paragraphs, *critical periods* in postnatal neurocognitive development define degrees of vulnerability, as reported by several authors following evidence from human and animal models (e.g., Rice and Barone 2000). In this regard, among humans, neurodevelopmental critical periods have different time windows for each primary sensory area: somatosensory (S1), visual (V1), and auditory (A1) (Pedrosa et al. 2022).

The above physiological, interactive considerations that condition brain and mental development are coupled to an additional variable linked to the host-gut microbiome, as will be commented on in the following paragraphs.

*

Microbiome and Brain/Mental Development

According to Greenhough et al. (2020),

> The human microbiome is an important emergent area of cross, multi and transdisciplinary study. The complexity of this topic leads to conflicting narratives and regulatory challenges... The social sciences and the humanities have begun to explore the microbiome as an object of empirical study and as an opportunity for theoretical innovation. They can play an important role in facilitating the development of research that is socially relevant, that incorporates cultural norms and expectations around microbes and that investigates how social and biological lives intersect.

As mentioned, the prenatal and postnatal – including infantile – periods in primate brain development represent critical periods characterised by rapid neuro-behavioural changes, during which environmental factors could have a long-term impact on the brain and behaviour. These events are time-coupled with postnatal gut microbiome installation and development. Given the significant health consequences associated with microbial perturbations, it has been proposed that the first one to 24 months of life represent a critical developmental window for establishing the gut microbiome (Hollister et al. 2015). Perturbations of these interactions may contribute to neurodevelopmental disorders (Borre et al. 2014). In this regard, Streit et al. (2021) reported that gut microbiome profiles are associated with cognitive performance measured with the Wechsler Preschool and Primary Scale of Intelligence III in 45-month-old children, controlling for gestational length, birth weight, congenital disease, or chromosomal abnormality.

As discussed by Ismail et al. (2017), under normal conditions, neuroplasticity exhibits a heterochronic cortex-specific developmental profile and is heightened during *critical and sensitive periods* of pre- and postnatal brain development. This allows for modelling experience-dependent structural and functional brain connections. Interestingly, relative to social settings, Turrell et al. (2002) considered that socioeconomic conditions across all life course stages appear to make unique contributions to cognitive function in late middle age. Their experience showed a strong, graded association between cumulative socioeconomic disadvantages and cognitive function. Those who occupied a low socioeconomic position during childhood and adulthood scored lower on every test than those who always occupied a high position. This statement would acquire additional relative weight, considering that developmental individual differences in neurocognitive profiles and host-microbiome interaction could bias individual socioeconomic position. Swann et al. (2017) demonstrated that the gut microbiome can imprint on the metabolic profile of the hippocampus and the frontal cortex by applying high-resolution ^{1}H NMR (^{1}H Nuclear Magnetic Resonance) spectroscopy in rodents.

> Given the importance of the microbiome for human health, both the stability and the response to disturbance of this microbial ecosystem are crucial issues.
>
> *(Relman 2012)*

At the human level, in terms of the host gut microbiome, according to Wopereis et al. (2014), gut microbial development in the neonatal period is influenced by several early-life factors. Diet, mainly, would drive further diversification towards an adult complexity, which would be reached around three years of age. As stated by the authors, this postnatal colonisation process provides signals known as microbe-associated molecular patterns that affect the maturation of the immune system. In this regard, induction of oral ingestive tolerance would be related to early maternal feeding involving secretory immunoglobulins A and G, together with the developing microbiota, essential for initiating an infant's secretory immunoglobulin A. Additionally, due to the capacity of intestinal microbial communities to regulate nutritional energy harvest, the gut microbiota may play a regulatory role in neurodevelopment during the first 1000 days (Robertson et al. 2019). According to these authors, despite under-five mortality rates falling by half since 1990, undernutrition continues to underlie 45% of all child deaths globally, and one-quarter of under-five children are stunted. As commented by Durda-Masny et al. (2022), the intestinal microbiota is involved in a wide range of processes that regulate child growth.

Prescott et al. (2016) stated that,

> Through the first 1000 days of life outside the womb, the stability and diversity of intestinal microbiota are increased, taking on an adult-like community structure. However, there are distinct compositional and functional characteristics to the preadolescent gut microbiome, and these reflect a supportive role in rapid growth and development.

Pre- and postnatal developmental stages usually interact with socioeconomic conditions in addition to the interacting host-microbiome development. In this regard, Colombo and Lipina (2005), Colombo (2007, 2019, 2020, 2022), and Lipina et al. (2005, 2009, 2013) reviewed and discussed pre- and early postnatal socioeconomic conditions affecting brain and behavioural development, besides its impact on feeding habits and resources.

Within this domain, it should be stressed that poverty interferes with optimal, balanced feeding during developmental years, which would combine and interact with modified development of host-microbiome interaction. As stated by Harrison and Taren (2018), income inequality is an underlying factor for the maladaptive changes in the microbiota in specific populations. The authors propose that this contributes to the health disparities observed between lower-income and higher-income populations in high-income countries. Hence, a central issue is whether inadequate developmental host-microbiome interacting conditions contribute to defining physical (pathology-prone) cognitive profiles and affect social- and behavioural profile drives. This is due to a growing body of evidence showing that microbes and symbionts can mediate behavioural changes in their hosts. According to Iddrisu et al. (2021), the gut microbiota of children who became malnourished in their first three years was poorly matured (in terms of microbial diversity) and different to that of healthy children.

The question raised by Prescott et al. (2016) of whether the risks of altered microbiota are equal among all socioeconomic groups was considered in studies by Miller et al. (2016). These involved associating socioeconomic status with gut microbiome alterations based on USA samples. It revealed that neighbourhood socioeconomic status explained 12–18% of the within-sample variability of colonic microbiota. As neighbourhood socioeconomic status increased, so did the within-sample diversity of the colonic sigmoid mucosa and faecal microbiota. Additionally, residence in neighbourhoods of higher socioeconomic status was associated with a greater abundance of *Bacteroides* (normally commensal, involved in complex molecules processing) and a lower abundance of *Prevotella* (normally associated with diets rich in plant-based consumption), suggesting that diet potentially underlies differences in gut microbiota composition. Gatya et al. (2022) reported gut microbiota alterations in undernourished children from Indonesia associated with diet and sociodemographic factors. These findings suggest the presence of socioeconomic variations in colonic microbiota diversity. Although statistically significant, the associations observed were generally modest, with neighbourhood socioeconomic status explaining up to 10–22% of the person-to-person variation in diversity indices. These figures, though, projected onto public demography imply the degradation of individual developmental conditions of significant numbers within a community.

Early Malnutrition

> Comparison of child growth patterns in 54 countries with WHO standards shows that growth faltering in early childhood is even more pronounced than suggested by previous analyses based on the National Center for Health Statistics reference.
> *(Victora et al. 2010)*

Leonard et al. (2003) state that the metabolic demands of relatively large brains during infancy are met by high levels of body fat reserves during the first year of life, in addition to a synergistic development of the gut microbiome to synthesise needed intermediate factors involved in neural cell regulation and growth. Hanson et al. (2013) analysed repeated measures of brain development of children between five months and four years of age from economically diverse backgrounds (n = 77). The authors found lower volumes of grey matter in frontal and parietal lobes in infants from low-income families. It is noted that the frontal lobes carry central processing of executive functions (planning, impulse control, attention control).

Let us consider body growth as a reference for normal and health-compromised child growth. According to the WHO Multicentre Growth Reference Study Group (2006),[1] a child is considered "stunted" whenever height exceeds two standard deviations below the WHO standards. Among the factors promoting stunting are maternal undernutrition, anaemia, tobacco use, infections, and indoor air pollution. In a report from 2018, several international institutions (United Nations Children's Fund/World Health Organization/International Bank for Reconstruction and Development/The World Bank) reported that globally in 2017, 150.8 million (22.2%) children under five years of age were stunted, with wasting threatening the lives of 7.5% (50.5 million) of children. Africa and Asia were the continents most affected by such malnutrition, bearing 39% and 55% (respectively) of global stunting prevalence in children under five years. As stated later (2019) in a joint report by those institutions, according to the anthropometric index, undernutrition consists of stunting (height/age +/- two standard deviations) and wasting (weight/age +/- two standard deviations).

Undernutrition has been a part of human experience throughout history and still accounts for 45% of deaths in children under five (WHO 2020). A report from 2021 estimated that 149.2 million children are stunted, and 13.6 million undergo malnutrition (UNICEF; WHO; World Bank Group 2021). Undernourished children tend to undergo a condition characterised by alteration of the gut microbiota composition (dysbiosis) (Hoffman et al. 2017), causing a decrease in immune system activity. According to Black et al. (2013), most stunting occurs in low- and middle-income countries in the first two years of life. Of clinical and public policy interest is the report quoted in Iddrisu et al. (2021), which states that specific subclinical alterations in the gut microbiome due to poor sanitary conditions and chronic exposure to environmental pathogens can lead to stunting even in the absence of apparent infections such as diarrhoea

Barrat et al. (2022) called attention to 45% of deaths among children under five years of age associated with undernutrition. Globally, almost 200 million children exhibit one of the two major forms of undernutrition – wasting (low weight-for-height) or stunting (low height-for-age) – and many of them are affected by both conditions. According to these authors, undernutrition is not due to food insecurity alone since growing evidence indicates that altered postnatal gut microbiome development would contribute to its pathogenesis.

In 1999, Mendez and Adair stated that after multivariate adjustment on tests run among more than 2000 Filipino children, severe stunting at age two remained significantly associated with later deficits in cognitive ability. The timing of stunting was also related to test performance, mainly because children stunted very early and tended to be severely cognitively affected. Studies performed by Chang et al. (2002) on children who were stunted at nine to 24 months of age and had taken part in a two-year intervention programme of psychosocial stimulation with or without nutritional supplementation were re-examined at age 11–12 years and compared with non-stunted children from the same neighbourhoods. The stunted group had more behavioural alterations as rated by their parents. They also had significantly lower arithmetic, spelling, word reading, and reading comprehension scores than the non-stunted children. In a follow-up study, Berkman et al. (2002) also observed similar effects on cognitive function at nine years. Guerrant et al. (2012) reported that many cognitive outcomes in early childhood illness studies reflect the multifactorial origin of developmental delay, including birthweight, household stimulation, and maternal behaviour.

Additionally, frequent diarrhoea during the first two years of life has been negatively correlated with cognitive development and early school performance. This impairment due to early diarrhoea was observed in verbal fluency (in semantic but not phonetic fluency) in Brazilian children by Patrick et al. (2005). Guerrant et al. (2012) considered that diarrhoea in children from impoverished areas during their first two years might cause, on average, an eight cm growth shortfall and ten IQ point decrement by the time they are seven to nine years old.

Following studies based on children cohorts from Brazil, Guatemala, India, the Philippines, and South Africa, Victora et al. (2008) concluded that poor foetal growth or stunting in the first two years of life leads to irreversible damage, including shorter adult height, lower attained schooling, reduced adult income, and decreased offspring birth weight. The authors added that damage suffered in early life leads to permanent impairment and might also affect future generations. Stunting between 12 and 36 months of age is associated with poorer cognitive performance and lower school achievement in middle childhood (Grantham-McGregor 2007). These authors reported that of 559 million children under five in developing countries, 156 million suffered from stunting, and 126 million lived in absolute poverty. A meta-analysis of the datasets showed that 43% of children below the poverty line who were stunted continued suffering from growth potential to around 40 months of age, though most children suffered from stunted growth through adulthood. In addition, the authors noted that poverty was associated with inadequate food and poor sanitation and hygiene, leading to increased infections and persistent stunting, poor maternal education, increased maternal stress and depression, inadequate stimulation at home, and reduced education enrolment. It is to be noted that according to the World Health Organization (WHO) (2021), 22% of all children under five were stunted in 2020, and almost 14 million children under five were affected by severe wasting. It should be stressed that these early conditions impinge upon educational performance and later social and employment insertion.

As stated by Kane et al. (2015),

> Malnutrition is a significant pediatric health problem worldwide, resulting in nearly half (45%) of all deaths (~3.1 million) in children younger than 5. Those who survive frequently suffer from long-term sequelae including growth failure and neurodevelopmental impairment. Although poverty, with its associated food insecurity, is a major risk factor for malnutrition.

Furthermore, Thompson et al. (2021) commented that,

> Children with wasting (low weight for height), a distinct but often overlapping manifestation of undernutrition, also face long-term health defects, which persist despite therapeutic refeeding... probably involving gut dysfunction compromising nutrient availability, uptake and use required for healthy growth.

Nahar et al. (2012) studied the impact of food and social stimulation on malnourished children from Bangladesh aged six to 24 months old. Child development was assessed using a revised version of Bayley Scales at baseline and after three and six months of intervention. None of the individual treatment groups showed a significant benefit in development or growth and some adherence to the follow-up. Despite some weight improvement in the combined treated group, all remained severely underweight. Ashraf et al. (2012) observed similar results in a follow-up study. Following observations quoted in Prentice et al. (2013) regarding key findings based on Victora et al. (2010) reports that in the poorer regions of Southeast Asia and Africa, height-for-age z scores start with a deficit at birth and decline further in the first two years of life before reaching an apparent plateau at age five. As stated in Ashraf et al. (2012), this finding prompted focused attention between nine to 24 months of age as a window of opportunity for interventions against stunting, targeting investments at the first 1000 days. Yet, as stated in Goyal et al. (2015),

> A common practice in understanding the origins, effects, and effectiveness of various treatment/prevention strategies for childhood undernutrition has been to focus on the "first 1,000 d," which begins with conception and ends approximately 2 y after birth (Prentice et al., 2013). Here, we call for an expanded view beyond these 1,000 d, particularly in relation to the brain, and a broader cellular, metabolic, and genetic view of our developmental biology that encompasses our gut microbial community (microbiota) and its genes (microbiome).

As posed by these authors, chronic undernutrition results in impaired cognitive abilities that are often not evident until the second or third decade, including effects on behaviour. Thus, continuous stunting reduces the window and the opportunity for physical and cognitive development. Given that brain organisation is under continuous remodelling based on experience at later ages, though with comparatively lesser metabolic demands, the brain may be vulnerable to poor nutrition during its initial development and ongoing remodelling. As early as 1986, Falkner debated the interaction of body composition and energy needs during growth. According to this author, the basic components of energy requirement are basal metabolic rate (BMR), energy spent in physical activity and in response to cold, and energy used in growth. The most characteristic feature of this energy requirement is a decline in per kilogram body weight from more than 100 kcal/kg per day to less than 45 kcal/kg per day. Much of this decline is caused by a decrease in BMR per kilogram secondary to a differential in the growth rate of organs with a high resting metabolic rate, e.g., the brain. The observations made by Leonard and Robertson (1992, 1994) and Leonard et al. (2003) regarding glucose consumption must be considered within this domain. This, in addition to the fact that the adult brain represents 2% of the body weight yet consumes ⊠20% of the body's energy, and by the end of the first 1,000 days, the brain represents 75% of its adult size (cf., Goyal et al. 2015). Metabolic demands of the brain are greater in childhood, as reported in Raichle (2010), brain glucose metabolism reaches adult levels by age two. It is twice that adult level by about age nine, and it returns to adult levels by the early 20's. This trajectory parallels the proliferation of synapses in the brain and their eventual pruning as the adult brain is sculpted. At approximately age 10, a child's brain represents 5–10% of body mass, consumes twice the glucose and 1.5 times the oxygen per gram of tissue compared with an adult's brain, and accounts for up to 50% of the total basal metabolic rate of the body (Chugani 1998). Results obtained by Kuzawa et al. (2014), based on positron emission tomography (PET), showed that,

> … glucose uptake in the cerebral cortex is more than twice as high during early- to mid-childhood than in adulthood. This dynamism reflects the additional energetic costs associated with over proliferation of neuronal processes and synapses before activity-dependent pruning in late childhood and adolescence, along with aerobic glycolysis, which is thought to rise in support of synaptic growth.

These authors support the concept that the rate of glucose uptake by the human brain, in both absolute terms and relative to the body's metabolic expenditure, does not peak at birth when the size of the brain relative to the body is largest, but in early childhood when synaptic densities and related metabolic processes are maximal.

*

According to Goyal et al. (2015),

> … children with undernutrition revealed that they harbor gut microbiota with delayed development: i.e., their microbiota have configurations that are younger than that of chronologically age-matched individuals who manifest healthy growth phenotypes… Moreover, the gut microbiota of undernourished children are not durably repaired with current therapeutic interventions; they revert to immature configurations after cessation of treatment. In other words, these individuals have a persistent developmental abnormality.

According to Million et al. (2017), early depletion in gut *Bifidobacterium longum*, a maternal probiotic known to inhibit pathogens, represents the first step in gut microbiota alteration associated with severe acute malnutrition. Later, the absence of the Healthy Mature Anaerobic Gut Microbiota leads to deficiencies, such as energy harvest, vitamin biosynthesis and immune protection, and is associated with diarrhoea, malabsorption and systemic invasion by microbial pathogens. As stated by de Clercq et al. (2016), the gut microbiota is involved in regulating food intake. This affects hormones that influence metabolic function and brain areas associated with eating behaviour, as Kairupan et al. (2016) reported. This microbiota-gut-brain axis represents a bidirectional signalling axis that regulates body weight by balancing appetite, storage, and energy expenditure.

Regarding cognitive recovery, research suggests that sustained early enriched environments provide an opportunity for significant recovery. However, besides maternal and child malnutrition in low-income and middle-income countries triggering undernutrition and showing a growing problem with overweight and obesity, as thoroughly analysed and stated in Black et al. (2013), its impact on the gut microbiome adds another dimension to the problem. Factors other than neighbourhood socioeconomic status contribute to microbiota composition. Subramanian et al. (2014) described that malnourished children express persistent gut microbiota immaturity, as described in children from Bangladesh. Additionally, these authors considered whether microbiota immaturity associated with severe or moderate acute malnutrition maintained during and beyond childhood also underscores the need to determine its physiologic, metabolic, and immunologic consequences and how they might contribute to the associated morbidities and sequelae of malnutrition, including cognitive development. As offspring shift from being entirely dependent on their mothers for food to self-sufficient providers, they also undergo several metabolically demanding processes, such as brain growth, as mentioned above.

Therefore, the potential consequences of food shortages during this early period are significant (Amato 2016), and the metabolic risks are substantial, especially considering that most juveniles are less efficient foragers than adults, impacting gut microbiome development and stability. Consequently, characteristics of the gut microbiota are probably most important for offspring during early development, when the impact of low socioeconomic conditions and malnourishment compromise cognitive developmental patterns, and the degree of recovery. In this regard, specific issues on brain systems should be considered *critical* developmental periods characterised by their dependency on specific external or humoral signals and on the regulatory system involved. As mentioned above, Turkheimer et al. (2003) compared the source of IQ variance in impoverished and affluent families. Coincidentally, Hanscombe et al. (2012) observed that following the application of three different indices of socioeconomic status (SES) at eight ages from infancy through adolescence, the emerging pattern appears that shared experiences explain more of the variance in children's performance on IQ tests in more disadvantaged backgrounds, than genetic components. Thus, although the genetic influence on IQ is the same in lower-SES families, shared environmental influence appears greater, suggesting that family-based environmental interventions might be the most relevant. Brito and Noble (2014) remarked on the impact of SES on brain development and linguistic, social, and cognitive stimulation. The authors summarise its impact,

> SES is a multidimensional construct, combining objective factors such as an individual's (or parent's) education, occupation, and income (McLoyd, 1998). Neighborhood SES is also often considered (Leventhal and Brooks-Gunn, 2000), as are subjective measures of social status (Adler et al., 2000). In 2012, 46.5 million people in the United States (15%) lived below the official poverty line (United States Census Bureau, 2012) and numerous studies have reported socioeconomic disparities profoundly affecting physical health, mental well-being, and cognitive development (Anderson and Armstead, 1995; Brooks-Gunn and Duncan, 1997; McLoyd, 1998; Evans, 2006). In turn, SES accounts for approximately 20% of the variance in childhood IQ (Gottfried et al., 2003) and it has been estimated that by age five, chronic poverty is associated with a 6- to 13-point IQ reduction (Brooks-Gunn and Duncan, 1997; Smith et al., 1997).

As quoted by Dewey and Begum (2011), based on Black et al. (2008), stunting impacted one-third of children under five years of age in low-income and middle-income countries, for a total of 178 million children. These figures, though, must be updated based on statistics from the WHO (2021), according to which 22% of all children under five were stunted by 2020, and almost 14 million children under five were affected by severe wasting (WHO, The Global Health Observatory[2]). As stated in Black et al. (2013), malnutrition is a major cause of morbidity and mortality in children in low and middle-income countries and is estimated to have caused 3.1 million deaths or almost half (45%) of all child deaths in 2011.

Dinh et al. (2016) reported that the gut microbiota of stunted children was enriched in pro-inflammatory taxa, whereas that of non-stunted children was enriched in probiotic bacterial species.

> The largest number of stunted children in the world live in southern Asia [13]. In India, 48% of children under the age of 5 were estimated to be stunted in 2005–6 [23]. India had the highest numbers of low birth weight (LBW) deliveries (7.5 million) globally in 2010 [24]. A longitudinal birth cohort study of children conducted between 2002 and 2006 in an urban slum community in Vellore, south India, found that 61% of children were stunted by the age of 3 years [19], with LBW significantly associated with stunting at 3 years of age (OR 3.63, 95% CI 1.36–9.70).
>
> *(Dinh et al. 2016)*

According to Victora et al. (2021), stunting and wasting remain public health concerns in low-income countries, where 4.7% of children are simultaneously affected by both, a condition associated with a 4.8-times increase in mortality. This description is further completed by anaemia in pregnant mothers, with a global prevalence of 36.8% (Karami et al. 2022). The prevalence of anaemia in women remains high and unabated in many countries.

As Robertson et al. (2023) stated, stunting, or linear growth failure, arises from a network of underlying factors, including inadequate dietary quantity and quality. It would affect 22% of children under age five worldwide. The author adds,

> Stunting is associated with infectious morbidity, reduced childhood survival and impaired cognitive development. The lifelong impacts of poor growth contribute to an intergenerational cycle of stunting and impaired development, lower educational attainment, and reduced adult economic productivity[4].

These authors report the maturation of the early-life gut microbiome in a cohort of 335 children from rural Zimbabwe from one-18 months of age. In this analysis, microbiome functionality instead of the taxonomic composition of the gut microbiome predicts linear and ponderal growth.

The United Nations joint report publication (2021) involved the Food and Agriculture Organization of the United Nations (FAO), the International Fund for Agricultural Development (IFAD), the United Nations Children's Fund (UNICEF), the UN World Food Programme (WFP), and the World Health Organization (WHO), as published by the UN Report and stating that "global hunger numbers rose to as many as 828 million in 2021", and by UNICEF in several web sites.

- **49 million** people more than a year earlier and **150 million** more than in 2019. Additionally, the report informed that,
- After remaining relatively unchanged since 2015, the proportion of people affected by hunger increased in 2020 and continued to rise in 2021 to **9.8%** of the world population. This compares against **8%** in 2019 and **9.3%** in 2020.
- Around **2.3 billion** people globally (**29.3%**) were moderately or severely food insecure in 2021 – **350 million** more compared to before the outbreak of the COVID-19 pandemic.
- Nearly **924 million** people (**11.7%**) faced food insecurity at severe levels, an increase of **207 million** people in two years.
- The gender gap in global food insecurity continued to rise in 2021 – **31.9%** of women were moderately or severely food insecure, compared to **27.6%** of men – a gap of more than 4% points, compared with 3% points in 2020.
- Almost **3.1 billion** people could not afford a healthy diet in 2020, up **112 million** from 2019, reflecting the effects of inflation in consumer food prices stemming from the economic impacts of the COVID-19 pandemic and the measures put in place to contain it.
- An estimated **45 million** children under five suffer from wasting the deadliest form of malnutrition, which increases children's risk of death by up to 12 times.
- **149 million** children under five had stunted growth and development due to a chronic lack of essential nutrients in their diets, while **39 million** were overweight.
- Global progress is being made on breastfeeding, with nearly **44%** of infants under six months of age being exclusively breastfed in 2020. This still falls short of the **50%** target by 2030. Of great concern is that two in three children are not fed the minimum diverse diet they need to grow and develop to their full potential.
- Projections are that nearly **670 million** people (**8%** of the global population) will still face hunger by 2030, even if global economic recovery is considered. This is a similar number to 2015, when the goal of ending hunger, food insecurity, and malnutrition by the end of that decade was launched under the 2030 Agenda for Sustainable Development.

It should be added that according to FAO, UNICEF, WFP, and WHO, between 702 and 828 million people were affected by hunger in 2021. Additionally, according to the UNICEF press release report (2022) and United Nations News, the global hunger crisis is pushing one child every minute into severe malnutrition in 15 crisis-hit countries (primarily African countries).

Worldwide, malnutrition contributes to almost half of the deaths in children under five years of age, claiming the lives of over three million children annually. It must be added that a child might not be hungry but still malnourished.[3] As stated earlier in this chapter, the Joint Child Malnutrition Estimates (JME) released in 2023 reveal insufficient progress to reach the 2025 World Health Assembly (WHA) global nutrition targets.

Finally, as stated by the World Food Programme,

An expected 345.2 million people projected to be food insecure in 2023 – more than double the number in 2020. This constitutes a staggering rise of 200 million people compared to pre-COVID-19 pandemic levels. More than 900,000 people worldwide are fighting to survive in famine-like conditions.[4]

World hunger is rising, affecting nearly 10% of the global population.[5] From 2019 to 2022, the undernourished population grew by as many as 150 million. FAO, IFAD, UNICEF, WFP, and WHO (2023) reported that while global hunger rates stalled between 2021 and 2022, many places face deepening food crises. This report documents that in 2022, approximately 2.4 billion individuals, predominantly women and residents of rural areas, did not have consistent access to nutritious, safe, and sufficient food, and child malnutrition is still alarmingly high. In 2021, 22.3% (148.1 million) children suffered from stunted growth, 6.8% (45 million) suffered from severe malnutrition, and 5.6% (37 million) were overweight.

In summary, as the social conditions exposed by these reports imply – when basic feeding requirements are not provided in adequate quality and quantity – the impact on the gut microbiome as a nexus with healthy host-gut microbiome interactions is wholly disrupted in a sustained condition. This affects the general physiology and brain/mental development to degrees of stunting with difficulty for a full recovery. When sickness ensues, it furthers the imbalance in host-gut microbiome equilibrium. The global debt of future development of large populations represents perhaps the more profound alteration in the social fabric and individual futures that should represent a prime concern and action for global public and political organisations. The above conditions represent the extreme worldwide evidence of social inequality in our social construction and a lack of priority in public spending. It is not only a matter of physical and mental health but also of the option to fully express individuality, i.e., affecting freedom. In this regard, "freedom" is a concept eroded by biological and social conditions.

Social inequalities are evident based on many forms of inadequate nutrition in women and children, suggesting a key role of poverty and low education in these conditions (see also Colombo 2023). This reinforces the need for multisectoral actions to deter human degrading developmental conditions and accelerate the social and cognitive progress of these communities. The overall picture drawn by the reported conditions on socioeconomic inequalities between and within countries reveals that they remain largely unabated and continue to curtail the rate of progress at a global scale, widening the cultural and cognitive development gap between world population socioeconomic strata. While a minority discusses the advantages and risks of artificial intelligence and welcomes spending efforts to explore outer space, a large sector of the global population is undergoing unacceptable developmental conditions. Human history continues to be driven by stark differences in developmental conditions, affecting the world's social fabric.

Notes

1 https://www.who.int/publications/i/item/924154693X.
2 https://www.who.int/data/gho/data/themes/topics/joint-child-malnutrition-estima tes-unicef-who-wb.
3 https://www.children.org/global-poverty/global-poverty-facts/.
4 https://www.wfp.org/global-hunger-crisis#:~:text=number.
5 https://www.actionagainsthunger.org/the-hunger-crisis/ world-hunger-facts.

References

Amato, Katherine R. "Incorporating the Gut Microbiota into Models of Human and Non-Human Primate Ecology and Evolution." *American Journal of Physical Anthropology*, vol. 159, no. S61, 2016, pp. 196–215, doi:10.1002/ajpa.22908.

Ashraf, H., *et al.* "A Follow-Up Experience of 6 Months after Treatment of Children with Severe Acute Malnutrition in Dhaka, Bangladesh." *Journal of Tropical Pediatrics*, vol. 58, no. 4, 2012, pp. 253–257, doi:10.1093/tropej/fmr083.

Barratt, Michael J., *et al.* "Gut Microbiome Development and Childhood Undernutrition." *Cell Host & Microbe*, vol. 30, no. 5, 2022, pp. 617–626, doi:10.1016/j.chom.2022.04.002.

Berkman, Douglas S., *et al.* "Effects of Stunting, Diarrhoeal Disease, and Parasitic Infection During Infancy on Cognition in Late Childhood: A Follow-Up Study." *The Lancet*, vol. 359, no. 9306, 2002, pp. 564–571, doi:10.1016/s0140-6736(02)07744-9.

Black, Robert E., *et al.* "Maternal and Child Undernutrition: Global and Regional Exposures and Health Consequences." *The Lancet*, vol. 371, no. 9608, 2008, pp. 243–260, doi:10.1016/s0140-6736(07)61690-0.

Black, Robert E., *et al.* "Maternal and Child Undernutrition and Overweight in Low-Income and Middle-Income Countries." *The Lancet*, vol. 382, no. 9890, 2013, pp. 427–451, pubmed.ncbi.nlm.nih.gov/23746772/.

Borre, Yuliya E., *et al.* "Microbiota and Neurodevelopmental Windows: Implications for Brain Disorders." *Trends in Molecular Medicine*, vol. 20, no. 9, 2014, pp. 509–518, doi:10.1016/j.molmed.2014.05.002.

Brito, Natalie H., and Kimberly G. Noble. "Socioeconomic Status and Structural Brain Development." *Frontiers in Neuroscience*, vol. 8, 2014, www.ncbi.nlm.nih.gov/pmc/articles/PMC4155174/, doi:10.3389/fnins.2014.00276.

Chang, S. M., *et al.* "Early Childhood Stunting and Later Behaviour and School Achievement." *Journal of Child Psychology and Psychiatry*, vol. 43, no. 6, 2002, pp. 775–783, doi:10.1111/1469-7610.00088.

Chugani, Harry T. "A Critical Period of Brain Development: Studies of Cerebral Glucose Utilization with PET." *Preventive Medicine*, vol. 27, no. 2, 1998, pp. 184–188, www.sciencedirect.com/science/article/pii/S0091743598902742, doi:10.1006/pmed.1998.0274.

Colombo, Jorge A., and Sebastián Lipina. *Hacia Un Programa Público de Estimulación Cognitiva Infantil: Fundamentos, Metodos Y Resultados de Una Experiencia de Intervención Preescolar Controlada.* Buenos Aires; México: Ediciones Paidos, 2005.

Colombo, Jorge A., *et al.* "Cerebral Cortex Astroglia and the Brain of a Genius: A Propos of A. Einstein's." *Brain Research Reviews*, vol. 52, no. 2, 2006, pp. 257–263, doi:10.1016/j.brainresrev.2006.03.002.

Colombo, Jorge A. *Pobreza y desarrollo infantil. Una contribucion multidisciplinaria.* Buenos Aires: Ediciones Paidos, 2007, pp. 97–113.

Colombo, Jorge A. *Our Animal Condition and Social Construction.* New York: Nova Science Publishers Inc, 2019. eBook ISBN: 978-971-53615-53583.

Colombo, Jorge A. *Creativity, a Profile for Our Species: Social and Neurocognitive Issues.* Newcastle upon Tyne: Cambridge Scholars Publishing, 2020.

Colombo, Jorge A. *Dominance Behavior: An Evolutive and Comparative Perspective.* Cham: Springer International Publishing, 2022.

Colombo, Jorge A. *Evolution and the Human-Animal Drive to Conflict.* London: Routledge, 2023, doi:10.4324/9781003387695.

de Clercq, Nicolien C., *et al.* "Gut Microbiota in Obesity and Undernutrition." *Advances in Nutrition*, vol. 7, no. 6, 2016, pp. 1080–1089, www.ncbi.nlm.nih.gov/pubmed/28140325, doi:10.3945/an.116.012914.

Dewey, Kathryn G., and Khadija Begum. "Long-Term Consequences of Stunting in Early Life." *Maternal & Child Nutrition*, vol. 7, no. 3, 2011, pp. 5–18, doi:10.1111/j.1740-8709.2011.00349.x.

Dinh, Duy M., *et al.* "Longitudinal Analysis of the Intestinal Microbiota in Persistently Stunted Young Children in South India." *PLoS ONE*, vol. 11, no. 5, 2016, p. e0155405, doi:10.1371/journal.pone.0155405.

Durda-Masny, Magdalena, *et al.* "The Mediating Role of the Gut Microbiota in the Physical Growth of Children." *Life (Basel)*, vol. 12, no. 2, 2022, p. 152, doi:10.3390/life12020152.

Eloe-Fadrosh, Emiley A., and David A. Rasko. "The Human Microbiome: From Symbiosis to Pathogenesis." *Annual Review of Medicine*, vol. 64, no. 1, 2013, pp. 145–163, www.ncbi. nlm.nih.gov/pmc/articles/PMC3731629/, doi:10.1146/annurev-med-010312-133513.

FAO, IFAD, UNICEF, WFP and WHO. *The State of Food Security and Nutrition in the World 2023: Urbanization, agrifood systems transformation and healthy diets across the rural–urban continuum.* Rome: FAO, 2023, doi:10.4060/cc3017en.

Farah, Martha J. "The Neuroscience of Socioeconomic Status: Correlates, Causes, and Consequences." *Neuron*, vol. 96, no. 1, 2017, pp. 56–71, doi:10.1016/j.neuron.2017.08.034.

Gao, Wei, *et al.* "Functional Network Development during the First Year: Relative Sequence and Socioeconomic Correlations." *Cerebral Cortex*, vol. 25, no. 9, 2015, pp. 2919–2928, doi:10.1093/cercor/bhu088.

Gatya, Mifta, *et al.* "Gut Microbiota Composition in Undernourished Children Associated with Diet and Sociodemographic Factors: A Case–Control Study in Indonesia." *Microorganisms*, vol. 10, no. 9, 2022, p. 1748, doi:10.3390/microorganisms10091748.

Goyal, Manu S., *et al.* "Feeding the Brain and Nurturing the Mind: Linking Nutrition and the Gut Microbiota to Brain Development." *Proceedings of the National Academy of Sciences of the United States of America*, vol. 112, no. 46, 2015, pp. 14105–14112. doi:10.1073/pnas.1511465112.

Grantham-McGregor, Sally. "Early Child Development in Developing Countries." *The Lancet*, vol. 369, no. 9564, 2007, p. 824, doi:10.1016/s0140-6736(07)60404-8.

Greenhough, Beth, *et al.* "Setting the Agenda for Social Science Research on the Human Microbiome." *Palgrave Communications*, vol. 6, no. 1, 2020, doi:10.1057/s41599-020-0388-5.

Guerrant, Richard L., *et al.* "The Impoverished Gut – a Triple Burden of Diarrhoea, Stunting and Chronic Disease." *Nature Reviews Gastroenterology & Hepatology*, vol. 10, no. 4, 2012, pp. 220–229, doi:10.1038/nrgastro.2012.239.

Hanscombe, Ken B., *et al.* "Socioeconomic Status (SES) and Children's Intelligence (IQ): In a UK-Representative Sample SES Moderates the Environmental, Not Genetic, Effect on IQ." *PLoS ONE*, vol. 7, no. 2, 2012, p. e30320, doi:10.1371/journal.pone.0030320.

Hanson, Jamie L., *et al.* "Family Poverty Affects the Rate of Human Infant Brain Growth." *PLoS ONE*, vol. 8, no. 12, 2013, p. e80954, doi:10.1371/journal.pone.0080954.

Harrison, Christy A., and Douglas Taren. "How Poverty Affects Diet to Shape the Microbiota and Chronic Disease." *Nature Reviews Immunology*, vol. 18, no. 4, 2018, pp. 279–287, doi:10.1038/nri.2017.121.

Hoffman, Daniel J., *et al.* "Microbiome, Growth Retardation and Metabolism: Are They Related?" *Annals of Human Biology*, vol. 44, no. 3, 2017, pp. 201–207, doi:10.1080/03014460.2016.1267261.

Hollister, Emily B., *et al.* "Structure and Function of the Healthy Pre-Adolescent Pediatric Gut Microbiome." *Microbiome*, vol. 3, 2015, p. 36, pubmed.ncbi.nlm.nih.gov/26306392/, doi:10.1186/s40168-015-0101-x.

Iddrisu, Ishawu, *et al.* "Malnutrition and Gut Microbiota in Children." *Nutrients*, vol. 13, no. 8, 2021, p. 2727, www.mdpi.com/2072-6643/13/8/2727/htm, doi:10.3390/nu13082727.

Isaacs, Elizabeth B., *et al.* "The Effect of Early Human Diet on Caudate Volumes and IQ." *Pediatric Research*, vol. 63, no. 3, 2008, pp. 308–314, www.nature.com/articles/pr200860, doi:10.1203/pdr.0b013e318163a271.

Ismail, Fatima Yousif, *et al.* "Cerebral Plasticity: Windows of Opportunity in the Developing Brain." *European Journal of Paediatric Neurology*, vol. 21, no. 1, 2017, pp. 23–48, doi:10.1016/j.ejpn.2016.07.007.

Kairupan, Timothy Sean, *et al.* "Role of Gastrointestinal Hormones in Feeding Behavior and Obesity Treatment." *Journal of Gastroenterology*, vol. 51, no. 2, 2016, pp. 93–103, doi:10.1007/s00535-015-1118-4.

Kane, Anne V., et al. "Childhood Malnutrition and the Intestinal Microbiome." *Pediatric Research*, vol. 77, no. 1–2, 2015, pp. 256–262, doi:10.1038/pr.2014.179.

Karami, Mohammadmahdi, *et al.* "Global Prevalence of Anemia in Pregnant Women: A Comprehensive Systematic Review and Meta-Analysis." *Maternal and Child Health Journal*, vol. 26, no. 7, 2022, pp. 1473–1487, doi:10.1007/s10995-022-03450-1.

Kuzawa, Christopher W., *et al.* "Metabolic Costs and Evolutionary Implications of Human Brain Development." *Proceedings of the National Academy of Sciences*, vol. 111, no. 36, 2014, pp. 13010–13015, doi:10.1073/pnas.1323099111.

Leonard, William R., and Marcia L. Robertson. "Nutritional Requirements and Human Evolution: A Bioenergetics Model." *American Journal of Human Biology*, vol. 4, no. 2, 1992, pp. 179–195, https://www.semanticscholar.org/paper/Nutritional-requirements-and-human-evolution%3A-A-Leonard-Robertson/2f6a7a714302a58cf6758ed86e3733166d3b16b5, doi:10.1002/ajhb.1310040204.

Leonard, William R., and Marcia L. Robertson. "Evolutionary Perspectives on Human Nutrition: The Influence of Brain and Body Size on Diet and Metabolism." *American Journal of Human Biology*, vol. 6, no. 1, 1994, pp. 77–88, doi:10.1002/ajhb.1310060111.

Leonard, William R., *et al.* "Metabolic Correlates of Hominid Brain Evolution." *Comparative Biochemistry and Physiology Part A: Molecular & Integrative Physiology*, vol. 136, no. 1, 2003, pp. 5–15, doi:10.1016/s1095-6433(03)00132-6.

Lipina, Sebastián J., *et al.* "Performance on the a-not-b Task of Argentinean Infants from Unsatisfied and Satisfied Basic Needs Homes." *Revista Interamericana de Psicología/Interamerican Journal of Psychology*, vol. 39, no. 1, 2005, pp. 49–60, https://www.redalyc.org/articulo.oa?id=28439106.

Lipina, Sebastián J., and Jorge A. Colombo. *Poverty and Brain Development during Childhood: An Approach from Cognitive Psychology and Neuroscience*. Washington, DC: American Psychological Association, 2009.

Lipina, Sebastián J., *et al.* "Linking Childhood Poverty and Cognition: Environmental Mediators of Non-Verbal Executive Control in an Argentine Sample." *Developmental Science*, vol. 16, no. 5, 2013, pp. 697–707, doi:10.1111/desc.12080.

Luby, Joan, *et al.* "The Effects of Poverty on Childhood Brain Development." *JAMA Pediatrics*, vol. 167, no. 12, 2013, pp. 1135–1142, pubmed.ncbi.nlm.nih.gov/24165922/, doi:10.1001/jamapediatrics.2013.3139.

Meloni, Maurizio. "The Social Brain Meets the Reactive Genome: Neuroscience, Epigenetics and the New Social Biology." *Frontiers in Human Neuroscience*, vol. 8, 2014, doi:10.3389/fnhum.2014.00309.

Mendez, Michelle A., and Linda S. Adair. "Severity and Timing of Stunting in the First Two Years of Life Affect Performance on Cognitive Tests in Late Childhood." *The Journal of Nutrition*, vol. 129, no. 8, 1999, pp. 1555–1562, doi:10.1093/jn/129.8.1555.

Miller, Gregory E., *et al.* "Lower Neighborhood Socioeconomic Status Associated with Reduced Diversity of the Colonic Microbiota in Healthy Adults." *PLoS ONE*, vol. 11, no. 2, 2016, p. e0148952, doi:10.1371/journal.pone.0148952.

Million, Matthieu, *et al.* "Gut Microbiota and Malnutrition." *Microbial Pathogenesis*, vol. 106, 2017, pp. 127–138, doi:10.1016/j.micpath.2016.02.003.

Nahar, B., *et al.* "Effects of a Community-Based Approach of Food and Psychosocial Stimulation on Growth and Development of Severely Malnourished Children in Bangladesh: A Randomised Trial." *European Journal of Clinical Nutrition*, vol. 66, no. 6, 2012, pp. 701–709, doi:10.1038/ejcn.2012.13.

Nieto-Sampedro, Manuel. "Plasticidad Neural: Del Aprendizaje a La Reparación de Lesiones." *Arbor-Ciencia Pensamiento Y Cultura*, no. 602, 1996, pp. 89–126.

Patrick, Peter D., *et al.* "Limitations in Verbal Fluency Following Heavy Burdens of Early Childhood Diarrhea in Brazilian Shantytown Children." *Child Neuropsychology*, vol. 11, no. 3, 2005, pp. 233–244, doi:10.1080/092970490911252.

Pedrosa, Laís Resque Russo, *et al.* "Time Window of the Critical Period for Neuroplasticity in S1, V1, and A1 Sensory Areas of Small Rodents: A Systematic Review." *Frontiers in Neuroanatomy*, vol. 16, 2022, doi:10.3389/fnana.2022.763245.

Prescott, Susan L., *et al.* "Biodiversity, the Human Microbiome and Mental Health: Moving toward a New Clinical Ecology for the 21st Century?" *International Journal of Biodiversity*, vol. 2016, 2016, pp. 1–18, doi:10.1155/2016/2718275.

Prentice, Andrew M., *et al.* "Critical Windows for Nutritional Interventions against Stunting." *The American Journal of Clinical Nutrition*, vol. 97, no. 5, 2013, pp. 911–918, doi:10.3945/ajcn.112.052332.

Raichle, Marcus E. "Two Views of Brain Function." *Trends in Cognitive Sciences*, vol. 14, no. 4, 2010, pp. 180–190, www.ncbi.nlm.nih.gov/pubmed/20206576/, doi:10.1016/j.tics.2010.01.008.

Relman, David A. "The Human Microbiome: Ecosystem Resilience and Health." *Nutrition Reviews*, vol. 70, no. suppl. 1, 2012, pp. S2–S9, doi:10.1111/j.1753-4887.2012.00489.x.

Rice, D., and S. Barone. "Critical Periods of Vulnerability for the Developing Nervous System: Evidence from Humans and Animal Models." *Environmental Health Perspectives*, vol. 108, no. suppl. 3, 2000, pp. 511–533, doi:10.1289/ehp.00108s3511.

Robertson, Ruairi C., *et al.* "The Human Microbiome and Child Growth – First 1000 Days and Beyond." *Trends in Microbiology*, vol. 27, no. 2, 2019, pp. 131–147, www.cell.com/trends/microbiology/fulltext/S0966-842X(18)30204-X, doi:10.1016/j.tim.2018.09.008.

Robertson, Ruairi C., *et al.* "The Gut Microbiome and Early-Life Growth in a Population with High Prevalence of Stunting." *Nature Communications*, vol. 14, no. 1, 2023, p. 654, www.nature.com/articles/s41467-023-36135-6, doi:10.1038/s41467-023-36135-6.

Streit, Fabian, *et al.* "Microbiome Profiles Are Associated with Cognitive Functioning in 45-Month-Old Children." *Brain, Behavior, and Immunity*, vol. 98, 2021, pp. 151–160, doi:10.1016/j.bbi.2021.08.001.

Subramanian, Sathish, *et al.* "Persistent Gut Microbiota Immaturity in Malnourished Bangladeshi Children." *Nature*, vol. 510, no. 7505, 2014, pp. 417–421, doi:10.1038/nature13421.

Swann, Jonathan R., *et al.* "Application of ^{1} H NMR Spectroscopy to the Metabolic Phenotyping of Rodent Brain Extracts: A Metabonomic Study of Gut Microbial Influence on Host Brain Metabolism." *Journal of Pharmaceutical and Biomedical Analysis*, vol. 143, 2017, pp. 141–146, doi:10.1016/j.jpba.2017.05.040.

Thompson, Alex J., *et al.* "Understanding the Role of the Gut in Undernutrition: What Can Technology Tell Us?" *Gut*, vol. 70, no. 8, 2021, pp. 1580–1594, gut.bmj.com/content/70/8/1580, doi:10.1136/gutjnl-2020-323609.

Turkheimer, Eric, *et al.* "Socioeconomic Status Modifies Heritability of IQ in Young Children." *Psychological Science*, vol. 14, no. 6, 2003, pp. 623–628, doi:10.1046/j.0956-7976.2003.psci1475.x.

Turnbaugh, Peter J., *et al.* "The Human Microbiome Project." *Nature*, vol. 449, no. 7164, 2007, pp. 804–810, www.ncbi.nlm.nih.gov/pmc/articles/PMC3709439/, doi:10.1038/nature06244.

Turrell, G., *et al.* "Socioeconomic Position across the Lifecourse and Cognitive Function in Late Middle Age." *The Journals of Gerontology Series B: Psychological Sciences and Social Sciences*, vol. 57, no. 1, 2002, pp. S43–S51, doi:10.1093/geronb/57.1.s43.

UN Report: "Global Hunger Numbers Rose to as Many as 828 Million in 2021." www.unicef.org, 6 July 2022, www.unicef.org/press-releases/un-report-global-hunger-numbers-rose-many-828-million-2021.

UNICEF, and WHO. "Levels and Trends in Child Malnutrition: Key Findings of the 2019 Edition." *UNICEF*, 1 Apr. 2019, www.unicef.org/reports/joint-child-malnutrition-estimates-levels-and-trends-child-malnutrition-2019.

UNICEF; WHO; World Bank Group. *Joint Child Malnutrition Estimates. Levels and Trends in Child Malnutrition: Key Findings of the 2019 Edition*. Geneva: World Health Organization, 2021, pp. 51–78.

United Nations. "Food." www.un.org/en/global-issues/food. Accessed 6 May 2024.

United Nations Children's Fund, World Health Organization, World Bank Group. *Levels and trends in child malnutrition: Key findings of the 2018 Edition of the Joint Child Malnutrition Estimates*. Geneva: World Health Organization, 2018.

United Nations Children's Fund, World Health Organization, International Bank for Reconstruction and Development/The World Bank. *Levels and Trends in Child Malnutrition: Key Findings of the 2019 Edition of the Joint Child Malnutrition Estimates*. Geneva: World Health Organization, 2019.

United Nations Children's Fund, World Health Organization, International Bank for Reconstruction and Development/The World Bank. *Levels and Trends in Child Malnutrition: UNICEF / WHO / World Bank Group Joint Child Malnutrition Estimates: Key findings of the 2023 edition*. New York: UNICEF and WHO, 2023.

Victora, Cesar G., *et al.* "Maternal and Child Undernutrition: Consequences for Adult Health and Human Capital." *The Lancet*, vol. 371, no. 9609, 2008, pp. 340–357, doi:10.1016/s0140-6736(07)61692-4.

Victora, Cesar G., *et al.* "Worldwide Timing of Growth Faltering: Revisiting Implications for Interventions". *Pediatrics*, vol. 125, no. 3, 2010, pp. e473–480, doi:10.1542/peds.2009-1519.

Victora, Cesar G, *et al.* "Revisiting Maternal and Child Undernutrition in Low-Income and Middle-Income Countries: Variable Progress towards an Unfinished Agenda." *The Lancet*, vol. 397, no. 10282, 2021, pp. 1388–1399, doi:10.1016/s0140-6736(21)00394-9.

Walter, Jens, and Ruth Ley. "The Human Gut Microbiome: Ecology and Recent Evolutionary Changes." *Annual Review of Microbiology*, vol. 65, no. 1, 2011, pp. 411–429, doi:10.1146/annurev-micro-090110-102830.

Wopereis, Harm, *et al.* "The First Thousand Days - Intestinal Microbiology of Early Life: Establishing a Symbiosis." *Pediatric Allergy and Immunology*, vol. 25, no. 5, 2014, pp. 428–438, onlinelibrary.wiley.com/doi/full/10.1111/pai.12232, doi:10.1111/pai.12232.

World Health Organization. Malnutrition – Key facts, 1 March, 2021, https://www.who.int/news-room/fact-sheets/detail/malnutrition#

11

BEYOND HUMAN PRIDE

We live in a period of rapid loss of biodiversity. Plants and animals are heading to extinction at alarming rates: it is too early to know with certainty if the microbial diversity of the biosphere is also in decline. However, a recent review of studies that investigated the sensitivity of terrestrial microbial community composition to forcers of global change (nitrogen, phosphorus, potassium, and organic carbon amendments, temperature change) showed that in the majority of cases, microbial community composition was indeed sensitive to disturbance. Thus, loss of microbial diversity is a real possibility in the near future, at least in some biomes. Some have called for the preservation of microbial DNA from a range of environments thought to be at risk, and have begun to preserve microbial life threatened by anthropogenic disturbance.

(Ley et al. 2008a)

Recent research is revealing surprising roles for microbiomes in shaping behaviours across many animal taxa—shedding light on how behaviours from diet to social interactions affect the composition of host-associated microbial communities and how microbes in turn influence host behaviour in dramatic ways... Once host-microbe associations are established, microbes can influence host behaviour in ways that have far-reaching implications for host ecology and evolution.

(Ezenwa et al. 2012)

The gut microbiota, the trillions of microbes inhabiting the human intestine, is a complex ecological community that through its collective metabolic activities and host interactions, influences both normal physiology and disease susceptibilities.

(Lozupone et al. 2012)

DOI: 10.4324/9781032698380-11

Humans can now be viewed as multispecies organisms operating within an ecological theatre.

(Prescott et al. 2016)

*

Progress has been made in experimental and clinical domains regarding host-gut microbiome interactions and their role in human development and health. However, its impact on contributing to defining human phenotypes remains elusive due to its multi-variable construct and the need to screen for age and cultural imprint. The impact of malnutrition on the development of human host-gut microbiota interactions is largely unexplored, contrary to its effects on the neuro-developmental process. On social grounds involving child development, the impact of malnutrition due to severe nutrient imbalances is only one aspect that claims urgent remedy. The above leads to an additional domain: poverty's impact on a child's mental and physical development and its social insertion. Also, whether it is associated with gut microbiome imbalances that further affect developmental dynamics (Swann et al. 2020). In this regard, it seems adequate to include results of its impact on individual epigenetics, involving processes related to DNA methylation and chromatin modification.

> The prospect that the social environment might sculpt our genome through modifying the epigenome is intriguing and might provide an explanation for the well-established relationship between socioeconomic status and physical health. Recent data suggest that social exposure early in life could alter epigenetic pro-gramming, which remains stable throughout life [Weaver et al., 2004]... This suggests a loop through which social exposure affect epigenetic states. These epigenetic states in turn affect social behaviour as well as behavioural pathologies.
>
> *(Szyf et al. 2008)*

Though several domains regarding research in host-microbiome interactions remain debatable, available information coincides with its mutually influencing roles, thus involving critical insights into human behaviour. There is an ongoing scientific debate (see Mayer et al. 2014, 2015) on the interaction between host and gut microbiome and whether they constitute a unitary, integrated event to be incorporated into host phenotype hereditability estimates. However, sound arguments from the experimental and clinical domains suggest a significant role of the microbiome in several health and pathological domains of living organisms, including humans. Among them, the gut microbiota has been shown to interact with host cognition in numerous laboratory animal model studies, to the point of representing a paradigm shift in neurosciences (Mayer et al. 2014), affecting feeding as well as driving the evolution of the social brain (Stilling et al. 2014). As stated by Mayer et al. (2014), the characterisation of the gut microbiome has initiated a paradigm shift not only in medicine but also in the primary and clinical domains of neuroscience.

Scientific progress on this domain has reached a point to inspire complex interac-tions in the formulation of animal behaviour, particularly in human behaviour, as posed by Allen et al. (2017). These authors proposed an intriguing interaction among domains of human conscious and unconscious behavioural construction as they conceptually coalesce psychoanalytic and the brain–gut-microbiota axis. In their words, based on several reports,

> If some information has relevance to a person's goals, it is not simply processed at a formal, logical level (sometimes referred to as Type 2 processes in thinking; e.g., Evans, 2007) but also leads to a subjective appraisal that is driven by an emotional response that is in turn tied up with a bodily response to the information. This emotional response will interact with the gut as well as, for example, cardiovascular, respiratory, and hormonal systems.
>
> In sum, a psychology of the brain–gut–microbiome axis should utilise multiple levels of analysis to understand this axis, not only at a physiological and intrapersonal level but also at a social and cultural level.
>
> Indeed, there has been some debate over whether humans (and other animals) can be thought of not as autonomous agents but as biomolecular networks or "holobionts," with the host and microbial genomes of the holobiont collectively referred to as the "hologenome".
>
> *(Bordenstein & Theis 2015; Douglas & Werren 2016, Theis et al. 2016)*

Several reports address the impact of microbiome composition on diverse rodent behaviours, as reported by Allen et al. (2017) and numerous other authors – several of them mentioned in the present chapter. Thus, despite promising results from pre-clinical investigation, further research must be performed to define the scope of microbiome roles in human development and behaviour, as discussed in Amato (2016) and Amato et al. (2017). According to Allen et al. (2017), the genetic information gleaned from the human genome project in mapping the human genome had relatively limited success in explaining human psychology. This may lead to an underplaying of the role of genetics in human cognition, emotion, and behaviour. However, as mentioned before, most genes in the human body are not the genes of human cells but rather are the genes of the microscopic organisms that dwell within humans.

Besides the above considerations, it should be remembered that behaviour is the composite outcome of a series of basic domains that may have genetic representation and be involved in the phenotypic definition. Dominguez-Bello et al. (2019) considered that coevolution of host-microbiome has likely shaped evolving phenotypes in all life forms on this predominantly microbial planet. By coevolving with the host, the microbiome would have shaped phenotypes in our ancestral lineages that survived dramatic environmental changes affecting survival probability, food availability, and intruding into the phenotypic expression and mood behaviour.

Given these conditionings, how much is left for the evidence of our individual social freedom that feeds the pride that humans lift as an identity banner?

The impact of the microbiome on human development and behaviour must be placed within the conceptual range of *critical* and *sensitive* periods in brain maturation (as previously discussed and in this chapter), concepts that tend to establish either a persistent or modulatory effect on brain development. Within the domain of microbiome development, as quoted in Douglas-Escobar et al. (2013), in animal models, the early introduction of normal microbes in mice initially exposed to a germ-free environment normalised their behaviour compared to controls, but not if it occurred after many weeks in which case the animals failed to normalise their behaviour.

Under normal circumstances, the host-microbiome association provides a mutual self-convenience association to the point that its alteration results in various health alterations. Regardless of the previous debate and unsolved questions, the host-microbiome community should be viewed as an ecological community of organisms involving a broad range of interactions contributing to sustaining a biological, behavioural, interactive complex unit. They should be viewed as mutually interdependent, regardless of the mechanisms involved in their establishment and subsistence. They converge in the main concept that human beings – in fact, also plant and animal Kingdoms – depend on the biological coalition of host and microbiome. To place this concept on an evolutionary domain, multicellular macro-organisms depend on the coalitionary presence of a series of micro-organisms that evolved aeons before macrospecies did, as described in the initial chapters.

Gilbert et al. (2012) discussed that symbiosis has transformed the classical conception of an insular individuality. The latter cannot be fully understood unless it involves the integration of the symbiotic microbiome. During evolution, multicellular life emerged from unicellular life forms that remain the dominant life form on the planet and often exist in a symbiotic or parasitic relationship with multicellular life (Dinan et al., 2015). In summary, Douglas and Werren (2016) proposed that the ecology and evolution of host-microbiome systems should be considered under an ecologic domain, where the host-microbiome builds an ecological community. Also, as mentioned by Risely (2020), the host-associated core microbiome was originally coined to refer to common groups of microbes or genes that were likely to be particularly important for host biological function. As this author states, the host-adapted core, as introduced by Shapira (2016), consists of specific microbial taxa that perform a function or functions that increase host fitness, either consistently or under ecological contexts, and their maintenance within the host population is a product of (host) natural selection. As stated in Shapira (2016),

> ... the "hologenome" model considers the genomes of the host and its microbes as one unit under selection. Starting with the observation that all animals (or plants) host diverse symbionts, the model proposes that symbionts can be inherited, as be exchanged with the environment; that association between host and symbiont affects the fitness of both; and that variation in the hologenome can be brought about by changes in either the host or the microbiota genomes.

According to Goodrich et al. (2014), host genetics and the gut microbiome can influence metabolic phenotypes, while host genetics variations affect the composition of the human gut microbiome. This, though Rothschild et al. (2018) considered that human microbiome composition is dominated by environmental factors rather than host genetics. These authors reported that family relatives with no history of a shared household do not have similar microbiomes, whereas microbiome similarity was observed among genetically unrelated individuals who share a household.

The mentioned dynamic, mutual dependency, exceeds the binary concept of health disease, as Shoubridge et al. (2022) discussed, to invade more subtle domains in constructing the human phenotype, mood, and social interactions. As considered by Münger et al. (2018), social interactions could strongly influence gut microbiome composition, particularly for individuals living in large social groups and spending ample time in social interactions, as Amato et al. (2017) described. For example, on primate comparative grounds chimpanzees acquire throughout their lifetime most of their gut phylotypes horizontally, i.e., through social interactions, rather than vertically from parent to offspring, as described in Moeller et al. (2016).

Finally, as posed by Prescott et al. (2016),

> The history of the genus Homo is a story of coevolution with microbes. For some 3 million years, commensal microbes have established their ecological niches on and within us.

The balanced quality of human feeding represents a significant item in the spectrum of physical and emotional factors affecting human development and behaviour. The previous pages provided evidence that human development and emotional and cognitive domains are not germane to the build-up of the human gut microbiome to the point that the host and gut microbiome conform to an interactive physiological entity.

The rates of social and health insecurities – most significantly among early ages – denounce critical flaws in our worldwide priorities and construction. They sustain or promote cognitive and educational developmental gaps among human communities in the face of extreme wealth exposed by comparatively minority social strata.

Thus, it could be concluded that both host genetics and environmental factors interact to mould host-gut microbiome profiles. This integrated, interactive perspective of human ecology is denounced in the WHO Report (2015). It supports the concept that reduced contact of people with the natural environment and biodiversity, and biodiversity loss in the broader environment, leads to reduced diversity in the human microbiota, which itself can lead to immune dysfunction and disease. Considering microbial diversity as an ecosystem service provider could generate a new paradigm in health security profiles. This would have an impact on human development sciences and public actions. As stated by Prescott et al. (2016), biodiversity loss affects ecosystem functioning, and significant disruptions of ecosystems can affect life-sustaining ecosystem goods and services, as well as quality feeding across cultural diversity, for humans are not foreign to, nor cannot evade its pertinence to an interactive biological construction. One that is based on the richness of cultural differences in our human world, which includes feeding habits as one biological source of such differences. A richness that resists cultural uniformity proposals endangers cultural diversity, human's innermost developmental potential. Thus, the species undergoes a growing bifront evolution, generating conditions for a gap of unpredictable consequences for the conception of an equal rights society.

References

Allen, Andrew P., *et al.* "A Psychology of the Human Brain–Gut–Microbiome Axis." *Social and Personality Psychology Compass*, vol. 11, no. 4, 2017, doi:10.1111/spc3.12309.

Amato, Katherine R. "Incorporating the Gut Microbiota into Models of Human and Non-Human Primate Ecology and Evolution." *American Journal of Physical Anthropology*, vol. 159, no. S61, 2016, pp. 196–215, doi:10.1002/ajpa.22908.

Amato, Katherine R., *et al.* "Patterns in Gut Microbiota Similarity Associated with Degree of Sociality among Sex Classes of a Neotropical Primate." *Microbial Ecology*, vol. 74, no. 1, 2017, pp. 250–258, doi:10.1007/s00248-017-0938-6.

Bordenstein, Seth R., and Kevin R. Theis. "Host Biology in Light of the Microbiome: Ten Principles of Holobionts and Hologenomes." *PLoS Biology*, vol. 13, no. 8, 2015, doi:10.1371/journal.pbio.1002226.

Dinan, Timothy G., *et al.* "Collective Unconscious: How Gut Microbes Shape Human Behavior." *Journal of Psychiatric Research*, vol. 63, 2015, pp. 1–9, doi:10.1016/j.jpsychires.2015.02.021.

Dominguez-Bello, Maria Gloria, *et al.* "Role of the Microbiome in Human Development." *Gut*, vol. 68, no. 6, 2019, pp. 1108–1114, gut.bmj.com/content/68/6/1108, doi:10.1136/gutjnl-2018-317503.

Douglas-Escobar, Martha, *et al.* "Effect of Intestinal Microbial Ecology on the Developing Brain." *JAMA Pediatrics*, vol. 167, no. 4, 2013, pp. 374–379, doi:10.1001/jamapediatrics.2013.497.

Douglas, Angela E., and John H. Werren. "Holes in the Hologenome: Why Host-Microbe Symbioses Are Not Holobionts." *MBio*, vol. 7, no. 2, 2016, doi:10.1128/mbio.02099-15.

Ezenwa, Vanessa O., *et al.* "Animal Behavior and the Microbiome." *Science*, vol. 338, no. 6104, 2012, pp. 198–199, doi:10.1126/science.1227412.

Gilbert, Scott F., *et al.* "A Symbiotic View of Life: We Have Never Been Individuals." *The Quarterly Review of Biology*, vol. 87, no. 4, 2012, pp. 325–341, doi:10.1086/668166.

Goodrich, Julia K., *et al.* "Human Genetics Shape the Gut Microbiome." *Cell*, vol. 159, no. 4, 2014, pp. 789–799, doi:10.1016/j.cell.2014.09.053.

Ley, Ruth E., *et al.* "Worlds within Worlds: Evolution of the Vertebrate Gut Microbiota." *Nature Reviews Microbiology*, vol. 6, no. 10, 2008, pp. 776–788, www.ncbi.nlm.nih.gov/pubmed/18794915, doi:10.1038/nrmicro1978.

Lozupone, Catherine A., *et al.* "Diversity, Stability and Resilience of the Human Gut Microbiota." *Nature*, vol. 489, no. 7415, 2012, pp. 220–230, www.ncbi.nlm.nih.gov/pmc/articles/PMC3577372/, doi:10.1038/nature11550.

Mayer, Emeran A., *et al.* "Gut Microbes and the Brain: Paradigm Shift in Neuroscience." *Journal of Neuroscience*, vol. 34, no. 46, 2014, pp. 15490–15496, www.jneurosci.org/content/34/46/15490.short, doi:10.1523/jneurosci.3299-14.2014.

Mayer, Emeran A. *et al.* "Gut/Brain Axis and the Microbiota." *The Journal of Clinical Investigation*, vol. 125, no. 3, 2015, pp. 926–938, doi:10.1172/JCI76304.

Moeller, Andrew H., *et al.* "Social Behavior Shapes the Chimpanzee Pan-Microbiome." *Science Advances*, vol. 2, no. 1, 2016, p. e1500997, advances.sciencemag.org/content/2/1/e1500997, doi:10.1126/sciadv.1500997.

Münger, Emmanuelle, *et al.* "Reciprocal Interactions between Gut Microbiota and Host Social Behavior." *Frontiers in Integrative Neuroscience*, vol. 12, 2018, doi:10.3389/fnint.2018.00021.

Prescott, Susan L., *et al.* "Biodiversity, the Human Microbiome and Mental Health: Moving toward a New Clinical Ecology for the 21st Century?" *International Journal of Biodiversity*, vol. 2016, 2016, pp. 1–18, doi:10.1155/2016/2718275.

Risely, Alice. "Applying the Core Microbiome to Understand Host-Microbe Systems." *Journal of Animal Ecology*, vol. 89, no. 7, 2020, pp. 1549–1558, doi:10.1111/1365-2656.13229.

Rothschild, Daphna, *et al.* "Environment Dominates over Host Genetics in Shaping Human Gut Microbiota." *Nature*, vol. 555, no. 7695, 2018, pp. 210–215, doi:10.1038/nature25973.

Shapira, Michael. "Gut Microbiotas and Host Evolution: Scaling up Symbiosis." *Trends in Ecology & Evolution*, vol. 31, no. 7, 2016, pp. 539–549, doi:10.1016/j.tree.2016.03.006.

Shoubridge, Andrew P., *et al.* "The Gut Microbiome and Mental Health: Advances in Research and Emerging Priorities." *Molecular Psychiatry*, vol. 27, 2022, doi:10.1038/s41380-022-01479-w.

Stilling, Roman M., *et al.* "Friends with Social Benefits: Host-Microbe Interactions as a Driver of Brain Evolution and Development?" *Frontiers in Cellular and Infection Microbiology*, vol. 4, no. 147, 2014, doi:10.3389/fcimb.2014.00147.

Swann, Jonathan R., *et al.* "Developmental Signatures of Microbiota-Derived Metabolites in the Mouse Brain." *Metabolites*, vol. 10, no. 5, 2020, p. 172, doi:10.3390/metabo10050172.

Szyf, Moshe, *et al.* "The Social Environment and the Epigenome." *Environmental and Molecular Mutagenesis*, vol. 49, no. 1, 2008, pp. 46–60, doi:10.1002/em.20357.

Theis, Kevin R., *et al.* "Getting the Hologenome Concept Right: An Eco-Evolutionary Framework for Hosts and Their Microbiomes." *MSystems*, vol. 1, no. 2, 2016, doi:10.1128/msystems.00028-16.

World Health Organization Press Release, Report on Health and Biodiversity Demonstrates Human Health Benefits from Protecting Biodiversity, 2015, https://www.unep.org/news-and-stories/press-release/report-health-and-biodiversity-demonstrates-human-health-benefits.

12

BRAIN AND GUT MICROBIOME

An Integrated View of Conditioning Developmental Processes

> Our species has shown us its dark side and thrown shadows over its bright profile through the ages. The obscure pages of our civilization's history continue to be written…
>
> *(Colombo 2022)*

It should be added that human civilisation's history continues to be written by chronic poverty indices, wars, dominance and oppression, torture, slavery, fanaticism, wealth concentration, and ecosystem degradation. However, this regretful profile continues beating creativeness, solidarity, and resistance – perhaps the basic values of our species.

> …even complex human behaviours reflect ancient mammalian neural systems that evolved to solve key problems in adaptive ways, with far-reaching consequences for even our most venerated human traits.
>
> *(Preston 2013)*

The development of new concepts and evidence involving brain-gut microbiome interactions expands over basic human tenets regarding health and differentiated cognitive developmental profiles. These affect socioeconomic developmental conditions, social behaviour, and cognitive performance. Thus, it enters the multivariable condition of human and ecosystem interactions and the likelihood of mental profile development. The impact on social programmes and policies is not negligible, nor should it be ignored; socioeconomic disadvantage is reproduced across generations. As stated by Scorza et al. (2019), (*epigenetic*) mechanisms act in the context of socioeconomic factors that cause families to remain in poverty, e.g., discrimination, economic conditions, educational and occupational opportunities, regional conditions, and family and community resources. It is an interactive sequence that, at the human level, requires further research to change the consequences on the behavioural profiles.

DOI: 10.4324/9781032698380-12

Our macrospecies dominant condition has progressively distorted our perception of our intimate relationship with the Natural Kingdom. This species-bound megalomania is not foreign to our evolutive cognitive and creative developmental capacities. This universal concept of dominance was incorporated into human cultural development and daily practice, with long-term consequences (Colombo 2022). Among them is the misleading intimate belief that the human species is detached from the natural eco-system interactions and dependence and can manipulate the ecosystem to its own convenience. In this cultural process, the human species' self-pride permeated collective cognitive levels, ignoring its integration and dependence on the ecosystem and progressively sliding into abusing the natural environment it has dwelled and interacted with throughout millennia.

This expanded cultural pride and sense of dominance also pressed the human species into disturbing and fracturing encounters with our species and environmental partners, which completes the integrated socio-ecological universal construction. The overall consequence has been the development and establishment of socio-political mis-adventures that projected the concept of a distorted relationship among social profiles, productive roles, and social rights onto cultural domains. This negation of an inte-grated society with the environment results – at another scale – in a disjointed concept of the human species within the natural- and social ecosystem. Ignoring this deformed concept that permeates our cultural constructions has consequences in understanding the mutual interdependence with the microbiome at one level and with the labour and productive forces at the human social level, i.e., two-prone source of potential conflicts.

*

> ... microbial genetic contributions are functional and include activities that influ-ence nutrient absorption, protection against pathogens, maintenance of barriers to the outside environment, and the manufacture of chemicals necessary for survival.
> *(Prescott et al. 2016)*

During the last decade or so, another biological domain has been explored that affects developmental conditions: the *gut microbiome* and its interaction with brain function and behaviour. Though most recorded experimental experience is based on experi-mental domains, accumulating reports based on human and non-human primates validate some comparative conclusions. These must be placed under the different neurodevelopmental characteristics involved as well as on the comparative brain/mental evolution between rodents and primates. This condition should justify further assessing the impact of host-gut microbiome interactions on humans, a matter pending completion in this developing field. From a biological standpoint, it cannot be ignored that, as a biological unit, humans extend beyond traditional microbiome-ignoring concepts. It would not be premature to consider human and primate species beha-viours as the result of an additional, dynamic, functional component, i.e., the gut microbiome.

The composite picture of these domains provides an integrated view of multiscale processes in human development. Neurobehavioural and gut microbiome domains traverse a developmental early labile period during which individual functional parameters could be environmentally affected. One main goal of this chapter is to place these developments within the general social frame of socio-economically compromised sectors that affect developing infant and juvenile populations and their social, creative insertion.

This integrated view would incorporate an additional perspective on the long-term developmental impact of cultural and nutrient dimensions, with its individual and social consequences. Unless humans take full conscience that their brain/mental capabilities took millennia to evolve as an integrated, mutually dependent sociobiological complex, humans will keep disjoining it from its creative and productive capacities. As described by Leonard and Robertson (1992, 1994) and Leonard et al. (2003), at the biological base of this shared unit, brain metabolism would account for 20–25% of resting energy demands in an adult human body, a proportion larger than the eight to ten % observed in other primate species, and still more than the three to five % allocated to the brain by other (non-primate) mammals. This representation of the brain underlines its metabolic demands – even larger (60–80%) during the first years of life – and the individual behavioural potential it provides. These developmental characteristics require satisfying the corresponding nutritional demands – an urgent, pending matter in human society, considering that human infants are born altricial (*relatively underdeveloped*). Unlike other primates, rapid brain growth continues into early postnatal life, as mentioned by Martin (1990) (cf. Leigh, 2012) and Rosenberg (1992). In this regard, according to Dewey and Begun (2011), long-term studies demonstrated that starting a nutritional intervention before three years of age has significant long-term physical effects, as well as on human capital and economic productivity in adulthood. It is also considered that nutritional supplementation of girls in early childhood significantly affects the body size of their offspring.

To provide energy stores for the metabolic demands of relatively large brains during infancy, humans have high body fat levels at birth and continue to gain fat during the first year of life. This process requires a healthy host-gut microbiome interaction. Due to the elevated energy demands during infancy, nutrient breast milk and weaning foods are essential to sustain the high brain and body growth rate characteristic of early life. According to Falkner and Tanner (1986), during infancy (with an estimated body weight of ten kg), brain metabolism would account for 60% + of the Resting Metabolic Rate (RMR), while in adulthood (with an estimated weight of 70 kg), brain metabolism would represent 20% of RMR.

As posed by Leonard (2012),

> … humans depart substantially from other primates in having a much higher-quality, more nutrient - dense diet than expected for a primate of our size. Adaptation to this dietary regime is reflected in our gut morphology, which in some aspects more closely resembles a carnivore than a folivore or frugivore. It appears that this dietary adaptation is a product of evolutionary trends for increased relative brain size (encephalization) in the hominin lineage over the past 2–2.5 million years. With a relatively larger brain size, the proportion of daily energy requirements used by the brain increases and necessitates a more easily digestible and energy-rich diet.

From the previous considerations emerges a fundamental social concern regarding the depth of the damage that current societies are incurring on large segments of the human population, notwithstanding other events during developmental stages. This implies the generation of a segregated vision of public policies of present societies, with citizens competing in less able conditions for their future, distorting humane values and compromising its future social collective development. A socioeconomic drive towards differential development is an immoral distribution of developmental resources (Colombo 2010) and is embedded in the behaviour of the *Homo* within the *sapiens*, as published earlier (Colombo 2021). It is a component of the prevailing bias in global development, subject to a basic behavioural premise of dominance behaviour (Colombo 2022; Colombo 2024).

*

An additional main conclusion from this chapter is our multilevel dependence on our actions and our true level of freedom – besides social interactions – of an organism embedded in many conditional variables (*Under Conditional Freedom*, Colombo 2013). These underwent an extended genetic exchange process among compatible species, adding a contingency to the host-gut microbiome domain and behavioural construction. From a human viewpoint, as recently stated by Chen and Ficetola (2020), progress in the collection and understanding of ancient DNA is progressively transforming our views on the evolution of the human species, mainly how an admixture with contemporary, archaic *hominins* through time has shaped the genomic construction of modern humans.

The basis of our nature emerged from an endless series of previous evolutionary events, which arose from adaptability and genetic potential under changing and often violent geo-climatic conditions that impinged on social behavioural construction, feeding habits and host-gut microbiome interactive composition. As mentioned before, anthropoids and related species of the *Homo* lineage that gave rise to *anatomically modern humans* probably had a chance for genetic exchange before speciation. In this ancestral trail from which *Homo sapiens* emerged, genetic certainties, ghostly remains, and different environmental conditions resulted in various human phenotypes. This statement does not limit itself to somatic events but includes individual and collective behavioural and neurocognitive profiles and drives. Numerous *Homo* species emerged from this species-ecological interaction, now extinct, with different characteristics and adaptive capacities to different ecological niches that possibly underwent genetic exchanges among proximate *Homo* variants with different gut microbiome universes.

From its harsh and complex species origin, humans have progressively attempted to construct foreseeable interactions and hence to design or model their peers through solid rules of education and their social environment. Today, sociopolitical hierarchies have the technology to address this goal through additional genetic and cultural means. What are its limits? What are the sources to balance such drifts in human potential? A large percentage of the population is immersed in stringent, unacceptable living conditions, under which the conditioning variables described here would complicate their ability and probabilities to reach full human creativity and survival potential.

Today, the process of cultural "homogenisation" – functional to the consumer-culture objective – gains strong allies in campaigns or advertising and financial pressures, which maintain a window of unnecessary or superfluous needs, where the essential or needed is confused with the banal or expendable, or the progress in bio-technological knowledge with destructive, consuming, campaigns deeply affecting environmental biodiversity. It is the mercantilist universe. The statistics in our time reflect the severe health, environmental, and economic deterioration, with a profound incidence on infant mortality and life expectancy in various communities, as well as on social exclusion and on the occupational index. These conditions amount to public policies of the worst kind of eugenics, publicly hidden as far as mass media awareness allows, in terms of generating public concern and action.

What once was the cradle of free-roaming and interacting australopithecines and *Homo* species developed into colonies and culturally oppressed, alienated populations of the African continent in the modern and contemporary world. Early European expansionism oppressed colonised and subdued African populations, not without help from co-terranean tribes. The ancestral, evolutive history of the early African emergence of *Homo* has been significantly replaced by a history of slavery and social and economic abuse of what became descendants of early migrant *Homo* tribes. Some consequences of dystopian developments have been previously discussed (Colombo 2007, 2015, 2019, 2021, 2022).

Excluded, *Homo sapiens pauper* wanders between submission, criminal violence, and revolutionary drives. On the one hand, the concept of the planet as a great "bowling alley" or a "kiosk" and its inhabitants as a malleable set of potential, virtual, or actual consumers; on the other, solidarity utopias with social inclusion.

The final, absolute value under the principles of the prevailing communities ("fittest"?), honed under socioeconomic strata, would be the species' survival regardless of the values it supports or the conditions in which it occurs. If so, then the concentration of power and the free-running scientific-technological progress – a powerful instrument of knowledge turned into an instrument of dominance – will continue its almost omnipotent development beyond any natural, social – collective – or ethical considerations. The fragmentation of the human community would then have undergone an irreversible change. The notion of community and humanity would be abandoned, replacing them with that of the market or privileged social strata, and individual differences transformed into oppression or abuse for group profiting purposes.

*

The Feeble Notion of Freedom: Freedom in a Conditioning Cage

As previously stated in Colombo (2013),

> … we must assume that we are biological entities and, therefore, belong to the Natural Kingdom, with creative potential and symbolic thought. The amazing thing about it is that such potential has emerged from the evolutionary dynamic -from the existing materials in nature-, in which we must recognize our fossils ancestors.

Implicit in this construction are human cognitive wealth, limitations, and ambivalences, the product of the friction between what has been characterised as *biological and cultural tectonic plates* (Colombo, 2010, 2021).

As explored in previous chapters, experience and integrative insights are accumulating geometrically on new data regarding the instalment and interactions between host and gut microbiome. It is the theoretical construct of a new integrated, dynamic, natural vision of human development and its microbiome under the variable socioeconomic circumstances of human civilisation. In this domain, what emerges is that the set of variables involved in physical and cognitive development has progressively incorporated conceptually new actors. Our traditional, classical view of humans and their circumstances must integrate a new domain with which living forms have shared since aeons past and with which they interact, as it does with its external environment. If the dimension of new technological advances and artificial intelligence (AI) is integrated into this vision, the latter should consider the shared interaction with gut microbiome conditioning.

Besides providing physiological clues on host-microbiome interactions of clinical significance, these advances would provide new perspectives to our cognitive development and dynamics considering the demanding socioeconomic and cultural conditions affecting neurodevelopment and characteristics of social insertion. This was also expressed as disparities in resting baseline EEG activity in the gamma frequency range in awake 6–9-month-olds from areas of East London with high socioeconomic deprivation (Tomalski et al. 2013), thus impacting the equal opportunities of the sociobiological fabric. Coincidently, based on surface-based morphometry on 3-T structural magnetic resonance imaging images on adult men from deprived neighbourhoods in Glasgow, Krishnadas et al. (2013) observed significantly smaller cortical surface area and cortical thickness in regions of the brain pertaining to language and executive function.

If our emotions and rational processes are affected by our gut microbiome, perhaps we should reconsider our historical proclamation on our freedom of thought and take a more careful stand on this issue (Colombo 2013). It would imply a formidable impact on human self-pride appraisal and feelings of freedom, should we take dear conscience that some bugs are involved in the process.

Humans feast on the self-generated idea that their degrees of freedom are solely constructed based on social and political interactive domains. Other domains hold the grip on our true freedom. They involve early social imprinting through family and educational structures, their binding to emotional clues, and, as recently described, their biological construction as an ecological unit of host-microbiome components.

> ... the bidirectional communication between the gut microbiome and the brain has emerged as a factor that influences immunity, metabolism, neurodevelopment and behaviour.
>
> *(Jašarević et al. 2016)*

The latter – as research reports described in this chapter would hold fully true for our species – implies sharing human behavioural construction with many microorganisms homing in its guts. In this respect, on evolutive grounds, one current theory on human species opportunity to increase brain mass and computational power is represented by the concept of an exchange with gut size, as compared with ruminant animals, based on the *expensive tissue hypothesis* (Aiello and Wheeler 1995; Roebroeks 2007). As expressed in their words,

> Our work complements that of Milton (1986, 1993; Milton and Demment, 1988), which suggests that the emergence of the hominids, and particularly of Homo, was associated with the incorporation of higher-quality foodstuffs into the diet. A high-quality diet was probably associated with a reduction in the size and therefore the energetic cost of the gut.
>
> *(Aiello and Wheeler 1995)*

This statement would seem to relate to the later evolving concept and further developments regarding early host-gut microbiome interactions among hominids. In this evolutive frame of thinking, a reduction in gut size would have been coupled with the growth of a new gut bacterial universe and changes in feeding habits and reformulating the rate of metabolic gain from food intake.

*

Concluding Remarks

> Despite our generally anthropocentric view of the world, it is the microbial population that dominates life on this planet in global diversity and in numbers. The human body itself serves as a scaffold for a multitude of bacteria, archaea, viruses, and eukaryotic microbes that inhabit discrete anatomical niches and outnumber our own somatic and germ cells by an order of magnitude (Turnbaugh et al, 2007).
>
> *(Eloe-Fadrosh and Rasko 2013)*

Here would seem to develop an evolutive contradiction as faced from the point of view of our human perspective. That is, the metabolic sharing with our gut microbiome has generated a release from the need to synthesise specific molecules and components, but at the same time, trapped our biological and humoral independence into a bacterial co-evolution. This, to the point that any significant break from this host-microbiome ecological construction derives into unnoticed or serious derangement of our full human brain capacity and health. Placing the issue into a more comprehensive view, the epigenetic construction of human personality and emotional character profiling, which relates closely to our most cherished behaviours – creativeness, solidarity – is not germane to an intimate interaction within our host-microbiome construction. Stilling et al. (2014) suggested a model in which the evolution of human sociability, which was accompanied by accelerated extension of the neocortex, is a key example of host-microbiome co-evolution and would be dependent on endosymbiotic developmental signals through the microbiota-gut-brain axis. Thus, the classical neurobehavioural concept based on a glial-neuronal frame of interactions, as implicit in the following statement, should incorporate input from the associated gut microbiome.

Behavioural and cognitive functions (unconscious and conscious), as well as their emotional components, are regulated from various neuronal groups located in brainstem centers... (Kandel, 2012).

(Gruart and Delgado-García 2023)

On developmental grounds, early (maternal-neonatal) life experience – to which feeding conditions and gut microbiome development should be added – affects the reward system, social hierarchy formation, and brain function, as mentioned by Ryakiotakis et al. (2023). This complex interaction involves social and nutritional variables, thus conditioning developmental and adulthood profiles.

*

In this theoretical framework, the grave question that ensues resides in where are left our drives for our most respected moral values, our caring for third parties, maternal drives, our sense and meaning of freedom if our true composition also

FIGURE 12.1 *"Attempt for Freedom"*. J.A. Colombo. Mixed technique, 2002.

involves the actions of an ecological host-gut microbiome unit? To what degree can we claim independence of our thinking processes? Could we regain such freedom or increase our degrees of freedom, perhaps a biological utopia? (Artistic representation in Figure 12.1) A quest whose display and success probability are conditioned by individual and social variables, as discussed before. Perhaps one target lies within reach, i.e., optimisation of rearing and developmental contexts that would improve our interactive, universal sharing of our present and future living conditions in less discriminatory social settings. To hold the concept that optimising brain/mental capabilities would slide the balance towards a more genuine self-conscience and individual degree of freedom, potentially releasing new creative ventures of human societies.

Is there any possible avenue to reconcile these unsettling considerations? To what extent is it feasible to override or condition our otherwise shared cognitive and emotional construction based on our host-gut microbiome interactions? Educational and self-enforcement values represent two conditions that may not be universally met related to developmental – physical, emotional, and social – conditions. This, considering that they merge fused with – conditioned by – dominant profiles of the social system construction and its ideological proposals.

Finally, plugging the gut microbiome into human personal emotional, cognitive, and physical stamina construction affects what could be a significant issue involving the motivations, personality profiles, and weaknesses of the main characters in our social, political, and cultural history. For, based on those that provide acceptable reasons to support the concept of a host-microbiome ecological unit, the history of humanity is a host-microbiome shared history. In other words, our existence would seem to lie beyond human pride, for we evolved from microorganisms and perform our actions and living intimately associated with bacteria and other members of the ancestral, ubiquitous, mutagenic specimens of Archean, prokaryotes, and eukaryotes, with whom we have shared profiles of our behaviour, frame of mind and drives since ever.

> The history of the genus Homo is a story of co-evolution with microbes. For some 3 million years, commensal microbes have established their ecological niches on and within us.
>
> *(Prescott et al. 2016)*

> Understanding why microbial diversity is necessary for the evolution and adaptation of the host, and why disease arises when such diversity is lost, is a fundamental question with still no definitive answer.
>
> *(Huitzil et al. 2018)*

Under a more general frame of thinking and considering human biological construction, how utopic is the human quest to increase individual freedom quota unless it provides the means and conditions to optimise brain development, increase mental power, and optimise freedom of thought?

References

Aiello, Leslie C., and Peter Wheeler. "The Expensive-Tissue Hypothesis: The Brain and the Digestive System in Human and Primate Evolution." *Current Anthropology*, vol. 36, no. 2, 1995, pp. 199–221, doi:10.1086/204350.

Chen, Wentao, and Gentile Francesco Ficetola. "Numerical Methods for Sedimentary-Ancient-DNA-Based Study on Past Biodiversity and Ecosystem Functioning." *Environmental DNA*, vol. 2, no. 2, 2020, pp. 115–129, doi:10.1002/edn3.79.

Colombo, Jorge A. *Pobreza y Desarrollo Infantil. Una Contribucion Multidisciplinaria*. Buenos Aires: Ediciones Paidós, 2007, pp. 97–113.

Colombo, Jorge A. *Somos La Especie Equivocada?: Una Mirada Evolutiva Sobre Las Responsabilidades De Nuestra Especie*. Buenos Aires: Eudeba, 2010.

Colombo, Jorge A. *Bajo Libertad Condicionada ¿Hacia La Conquista De Grados De Libertad?* Buenos Aires: Ediciones Imago Mundi, 2013.

Colombo, Jorge A. *Los Homo Sabios*. Buenos Aires: Buenos Aires Books, 2015.

Colombo, Jorge A. *Our Animal Condition and Social Construction*. New York: Nova Science Publishers Inc, 2019. eBook ISBN: 978-971-53615-53583.

Colombo, Jorge A. *The Homo within the Sapiens: (on the Animal-Driven Human Nature)*. New York: Nova Science Publishers, 2021.

Colombo, Jorge A. *Dominance Behavior: An Evolutive and Comparative Perspective*. Cham: Springer International Publishing, 2022.

Colombo, Jorge A. *Evolution and the Human-Animal Drive to Conflict: A Psychobiological Perspective*. Abingdon, Oxon: Routledge, 2024.

Dewey, Kathryn G., and Khadija Begum. "Long-Term Consequences of Stunting in Early Life." *Maternal & Child Nutrition*, vol. 7, no. 3, 2011, pp. 5–18, doi:10.1111/j.1740-8709.2011.00349.x.

Eloe-Fadrosh, Emiley A., and David A. Rasko. "The Human Microbiome: From Symbiosis to Pathogenesis." *Annual Review of Medicine*, vol. 64, no. 1, 2013, pp. 145–163, www.ncbi.nlm.nih.gov/pmc/articles/PMC3731629/, doi:10.1146/annurev-med-010312-133513.

Falkner, Frank T., and J. M. Tanner. *Human Growth: A Comprehensive Treatise*. New York: Plenum, 1986.

Gruart, Agnès, and José M. Delgado-García. "Neural Bases of Freedom and Responsibility." *Frontiers in Neural Circuits*, vol. 17, 2023, doi:10.3389/fncir.2023.1191996.

Huitzil, Saúl, *et al.* "Modeling the Role of the Microbiome in Evolution." *Frontiers in Physiology*, vol. 9, 2018, doi:10.3389/fphys.2018.01836.

Jašarević, Eldin, *et al.* "Sex Differences in the Gut Microbiome–Brain Axis across the Lifespan." *Philosophical Transactions of the Royal Society B: Biological Sciences*, vol. 371, no. 1688, 2016, p. 20150122, doi:10.1098/rstb.2015.0122.

Krishnadas, Rajeev, *et al.* "Socioeconomic Deprivation and Cortical Morphology." *Psychosomatic Medicine*, vol. 75, no. 7, 2013, pp. 616–623, doi:10.1097/psy.0b013e3182a151a7.

Leigh, Steven R. "Brain Size Growth and Life History in Human Evolution." *Evolutionary Biology*, vol. 39, no. 4, 2012, pp. 587–599, doi:10.1007/s11692-012-9168-5.

Leonard, William R., and Marcia L. Robertson. "Nutritional Requirements and Human Evolution: A Bioenergetics Model." *American Journal of Human Biology*, vol. 4, no. 2, 1992, pp. 179–195, https://www.semanticscholar.org/paper/Nutritional-requirements-and-human-evolution%3A-A-Leonard-Robertson/2f6a7a714302a58cf6758ed86e3733166d3b16b5, doi:10.1002/ajhb.1310040204.

Leonard, William R., and Marcia L. Robertson. "Evolutionary Perspectives on Human Nutrition: The Influence of Brain and Body Size on Diet and Metabolism." *American Journal of Human Biology*, vol. 6, no. 1, 1994, pp. 77–88, doi:10.1002/ajhb.1310060111.

Leonard, William R., *et al.* "Metabolic Correlates of Hominid Brain Evolution." *Comparative Biochemistry and Physiology Part A: Molecular & Integrative Physiology*, vol. 136, no. 1, 2003, pp. 5–15, doi:10.1016/s1095-6433(03)00132-6.

Leonard, William R. "Human Nutritional Evolution." In *Human Biology: An Evolutionary and Biocultural Perspective*, edited by Sara Stinson, Barry Bogin, and Dennis O'Rourke. New Jersey: Wiley, 2012, pp. 251–324, doi:10.1002/9781118108062.ch7.

Martin, R. D.Evolution of the Brain in Early Hominids. *Ossa*, vol. 14, 1989, pp. 49–62.

Prescott, Susan L., *et al.* "Biodiversity, the Human Microbiome and Mental Health: Moving toward a New Clinical Ecology for the 21st Century?" *International Journal of Biodiversity*, vol. 2016, 2016, pp. 1–18, doi:10.1155/2016/2718275.

Preston, Stephanie D. "The Origins of Altruism in Offspring Care." *Psychological Bulletin*, vol. 139, no. 6, 2013, pp. 1305–1341, doi:10.1037/a0031755.

Roebroeks, Wil. *Guts and Brains: An Integrative Approach to the Hominin Record*. Leiden: Leiden University Press, 2007.

Rosenberg, Karen R. "The Evolution of Modern Human Childbirth." *American Journal of Physical Anthropology*, vol. 35, no. S15, 1992, pp. 89–124, doi:10.1002/ajpa.1330350605.

Ryakiotakis, Ermis, *et al.* "Maternal Neglect Alters Reward-Anticipatory Behavior, Social Status Stability, and Reward Circuit Activation in Adult Male Rats." *Frontiers in Neuroscience*, vol. 17, 2023, doi:10.3389/fnins.2023.1201345.

Scorza, Pamela, *et al.* "Research Review: Intergenerational Transmission of Disadvantage: Epigenetics and Parents' Childhoods as the First Exposure." *Journal of Child Psychology and Psychiatry*, vol. 60, no. 2, 2019, pp. 119–132, doi:10.1111/jcpp.12877.

Stilling, Roman M., *et al.* "Friends with Social Benefits: Host-Microbe Interactions as a Driver of Brain Evolution and Development?" *Frontiers in Cellular and Infection Microbiology*, vol. 4, no. 147, 2014, doi:10.3389/fcimb.2014.00147.

Tomalski, Przemyslaw, *et al.* "Socioeconomic Status and Functional Brain Development – Associations in Early Infancy." *Developmental Science*, vol. 16, no. 5, 2013, pp. 676–687, onlinelibrary.wiley.com/doi/10.1111/desc.12079/abstract, doi:10.1111/desc.12079.

INDEX